The Love Story of Creation

The Love Story of Creation

* * * * * *

Book One

* * * * * *

The Creative Adventures of God, Quarkie, Photie, and Their Atom Friends

* * * * * *

Edward J. Ruetz

iUniverse, Inc.
New York   Bloomington

iUniverse books may be ordered through booksellers or by contacting:

*iUniverse
1663 Liberty Drive
Bloomington, IN 47403
www.iuniverse.com
1-800-Authors (1-800-288-4677)*

*Because of the dynamic nature of the Internet, any Web addresses or
links contained in this book may have changed since publication and may
no longer be valid. The views expressed in this work are solely those of
the author and do not necessarily reflect the views of the publisher, and
the publisher hereby disclaims any responsibility for them.*

*ISBN: 978-1-4401-8838-1 (sc)
ISBN: 978-1-4401-8840-4 (hc)
ISBN: 978-1-4401-8839-8 (ebook)*

Printed in the United States of America

iUniverse rev. date: 12/22/2009

Contents

Preface   vii

In Gratitude   xix

Characters   xxi

Chapters   xxv

1   Vision of the Quaternity   1
2   Creative Dawn   5
3   Photie and Quarkie, the First Creations   8
4   The Birthing of Neutrons, Electrons, and Neutrinos   16
5   The Red, Glowing Universe   20
6   The Creation of the Hydrogen Atom   25
7   A New Cocreative Mission Is Begun   30
8   The Nuclear Furnace Ignites   35
9   The Creativity of the Red Star and Birth of Carbon   40
10   Hydro's New Friend and the Creation of Oxygen   45
11   The Collapse of the Red Star   51
12   The Red Sun Explodes   55
13   Hydro Challenges His Atom Friends   62
14   Proud Hydro Looks to Take Leadership of the Creative Process   66
15   Hydro's Fatal Temptation   72
16   Phenomenal Photie's Tale   78
17   Beacons in a Vacuous Ocean of Space-time   87
18   The Divine Beings Give the Atom Characters a Guided Tour of a Spiral Galaxy   93
19   A Journey toward a Special Destiny   99
20   An Exploding Supernova—The Provoking Event   107
21   An Astounding Journey   114
22   Catastrophe Averted   119
23   The Atom Friends Land on Planet Earth   127
24   A New Creation Is Birthed   134
25   Prokye Is Introduced to His Cocreators   142
26   The Creation of Prokye's Image, a New Prokaryote Bacterium   149
27   Quarko's Suggestion Leads to a New Creation   157

28 Nitro and Hydrojean Meet a New Friend   165
29 Phosie and Photie Volunteer for a Creative Journey 171
30 Back to the Past: The Formation of the Solar Planets 177
31 Back to the Past: A Visit to the Four Rock Planets   183
32 Back to the Past: A Visit to the Five Gaseous Planets 196
33 Prokye 2 Grows toward Maturity   206
34 Photie and Prokye 3 Find a New Power Source   213
35 Phosie and Prokye 4 Work to Harness the Energy of the
    Sun   224
36 The Quark Trio Embarks on a New Mission   234
37 Back to the Past: A Journey through the Universe   241
38 Back to the Past: A Journey into and around Planet
    Earth   249
39 Back to the Past: A Return to Prokye 1's Birthplace 259
40 Prokye 2 Develops an Improved Replication System   268
41 Prokye 2's Surprise Reunion with Prokye 1   273
42 Prokye 3 Comes Home   282
43 Homecoming for the Crew of Prokye 4   293
44 Back to the Past: Dabar Leads the Six Atom Friends on
    a Journey   299
45 Back to the Past: Dabar and Friends Continue Their
    Journey   306
46 Divine Beings Dialogue with Their Atom Friends:
    Where Do We Go Here?   314

Epilogue   321

Afterword—The Future of Creation   323

Sources Consulted   343

Glossary   355

Preface

Fact, fantasy, faith, fiction! I have utilized all of these—literary genre, scientific fact, and religious faith—in order to write this story of creation. Coming from a Christian tradition, I have depicted God as the source of this creation story. I have written this account of the creation and evolution of the universe for a young adult audience who has a limited understanding of this scientific story of creation. I have written it in a simple but scientifically accurate manner. Fifteen years of presenting workshops on the scientific story of creation, using picture slides and PowerPoint, have been of great help in writing this book.

I am profoundly aware of the dangers involved in attempting to tell the scientific story of creation from a faith perspective. I have attempted this task because I believe that there is a great need for Christianity to come to grips with this creative tale, study it, and then assimilate its principles and truths into our traditional beliefs and practices.

I understand the sharp difference between science and faith. The scientist—astronomer, physicist, biologist, chemist, zoologist, anthropologist—is interested in discovering the facts about our universe and all of its community of beings, including us humans. They have developed rigorous methodologies to carry out their research in order to establish scientific fact and truth. From these experimental facts and truths, they develop theories to explain the creation and evolution of our universe and its community of beings. They are deeply interested in the question of origins. But the limits of their scientific methodologies do not allow them to study and theorize about a personal God as originator of the universe. They look exclusively to scientific methods in order to try to explain the origins of our magnificent universe and its diverse community of beings.

On the other hand, theologians are also interested in these same origins. But their methodologies differ greatly from the scientists. They also use rigorous methodologies to research their sacred scriptures, traditions, and history, whether they are Jewish, Confucianist, Hindu, Buddhist, Christian, Moslem, or a member of other major religious traditions. They approach their theological labors from a faith perspective. This faith is beyond the ability of science to test and verify experimentally with its technologies and methodologies. Nonetheless, the belief in a Higher Being, a God, Triune God, Allah, Jehovah, or any other such God or gods allows the believer to posit certain religious conclusions about the creation of the universe and its evolution. This approach allows theologians to arrive at religious truths concerning origins that are radically different in their essential nature from scientific fact, truth, and theory. Yet I deeply affirm that there is a profound connection between faith and science in studying origins of the universe and its community of beings.

Therefore, it was with profound trepidation that I began to write this book on the scientific story of creation with God as the originator of this stupendous creativity. Yet I believe that there is an urgent necessity for the believer, in my case, a Christian believer, to confront and understand science's story of how the universe and its community of beings came into existence and evolved over a period of 15 billion years. So I set out on this journey of writing this creation story in a simple, but forthright manner, with God as the originator of this enormous creativity. You will have to judge for yourself as to whether my effort is helpful to those of you who are trying to find your way between science and faith.

In recent years, some theologians, including Dr. John Haught, began to speak about "layered explanations" in relation to the story of evolution. Haught writes:

By layered explanations I mean to point out that most things in our experience admit of more than one level of explanation (possibly showing) how life may be explained simultaneously by both science and theology (Haught 2007, 141).

According to this approach, theologians study and observe the theory of the evolution of the universe and planet Earth and all of the facts and artifacts that give meaning to it. They do not invade and tamper with the scientific theory of evolution itself. They attempt to plum its story at a deep level.

The Divine Beings of faith, in this book, having set into motion the creation of the universe through their ecstatic energy of love, empower the first particles and subsequent atoms, photons, and molecules to create the universe. Through chance, accidents, randomness, contingency, natural selection, and deep time, the atom and photon families carry out their creative activity over the 15 billion years up to the present.

Theologians speak of "God's patience." As the originator of the universe, and after empowering the first particles and photons to carry out the creative process within it, the Godhead has been engaged in observing, but not tampering with, nature's creative work over these 15 billion years and will continue this patient activity into the future. Yet, at the time of the origins of biological life, when living beings came into existence and when Homo sapiens achieved consciousness, theologians see the Divine Beings engaged with and vivifying the original living beings and energizing Homo sapiens' consciousness at a deep level below the more surface levels of the physical, chemical, molecular, and environmental aspects of these forms of life.

Some scientists speculate that the universe will continue in a "steady state," ever expanding into the future. Others believe that the universe will end with it coming to a point when matter will reverse itself and crunch back onto the point from which it started. This is called the "big crunch theory."

Some theologians, using layered explanations and speculating from outside the evolutionary story, observe that in its processes, there are time and space; simplicity and complexity; unity and diversity; creation and extinction; suffering and resurrection; and novelty and beauty. They further speculate that our still unfinished universe will ultimately evolve through the coming eons to achieve unity and perfection. They point to Teilhard de Chardin's intuition that in the distant future, when the universe will have achieved this ultimate point, the Godhead will greet the creative beings who have achieved this unity at what he called "Point Omega" (deChardin 1955, 271f). At this ultimate point, the Godhead will greet the creative beings who have achieved this unity. Thus, in this theological speculation, the Godhead originated or gave birth to the universe, infusing it with the love energy used initially to birth it. Throughout the universe's 15-billion-year journey, the Godhead has patiently accompanied and energized its creators—the atoms, molecules, and photons—and their creations, especially its living beings, *at a deep level*, but *not* directing or determining their creative activities. Then, at Point Omega, the Godhead will receive them with love and relate to them with graciousness as it has finally achieved its unity and finished form.

In writing *The Love Story of Creation*, I have used a modified form of the ancient Jewish literary genre called midrash. In an Oxford dictionary, the editors write:

> Midrash, derived from the root *darash* (inquire), denotes the literature that interprets scripture in order to extract its full implications and meaning. Frequently the rabbis allowed themselves to indulge in far-fetched interpretations and narratives that was clearly intended to delight the fancy as well as to instruct (Werblowsky and Wigoder 1997, 463).

As you read this creation story, you will soon learn that I have indulged in "far-fetched interpretations and narratives" that are "clearly intended to delight the fancy of the reader, as well as instruct." My basic intent is to use narrative and fantasy, far-fetched as they may seem to be, to instruct you, the reader, about the scientific story of creation.

In telling this story, my greatest consternation and anxiety is my portrayal of the Creator God as a Quaternity and not as the traditional Trinity of Persons. I embrace the traditional belief of distinct persons but one God. I understand that I open myself to intense criticism at the hands of my co-religionist Catholics, from bishops to priests, theologians, and laypeople. But remember that I am using the literary genre midrash.

Let me attempt to explain my reason for using a Divine Quadernity rather than a Trinity. I have used divine names that come from the Hebrew and Christian scriptures. In the first chapter of the book of Genesis, Elohim is the Hebrew name given to the God who creates the universe in six days and rests on the seventh. "And Elohim said: 'Let there be light.' And light appeared" (Genesis 1, 3).

In the book of Wisdom (Catholic Christians accept it as one of the canonical books, while other Christians receive it as part of the apocryphal writings and not canonical), the writer personifies the Wisdom of God as being feminine. In Greek, the word for *wisdom* is *Sophia*. The Christian tradition has never accepted Wisdom or Sophia as a person of the Trinity. Wisdom is but an attribute of the one true God in this sapiential book. I depict Sophia as a person of God.

Since I chose to portray the first moment of creation not as a "big bang" but as a "birthing forth" of the universe, I decided to use a husband and wife, Elohim and Sophia, as the originators who give birth to this stupendous creativity in community with the other two Divine Persons in the Quaternity.

I have utilized the Greek word *Dabar* for the word the Son of God. This word appears in the first chapter of John's gospel. "In the beginning was the Word (Dabar), and the Word was with God and the Word was God. He was there with God. Through him all things were made, and without him, not one thing came into being" (John 1, 1–3). *Dabar* can also be translated as the personification of creative energy.

I again went back to the book of Genesis for a word to depict the Holy Spirit, the third Person of the Trinity in Christian theology. The biblical writer speaks of the chaos that existed at the beginning of creation. Hovering over this chaotic scene was *ruah*, the spirit, the breath, the wind (Gen. 1, 1–2). I have used the capitalized name Ruah for the Holy Spirit. I describe the mutual ecstatic love existing between Elohim, Sophia, and Dabar as being personified truly in the Person of Ruah, the Holy Spirit (Bruteau 1997, 27). In addition, I describe Ruah, who hovers over Elohim, Sophia, and Dabar, as a divine ghostly godmother. The Divine Beings decide to share their infinite energy of love with other beings. Their unlimited ecstatic love energy bursts forth into the blackness outside of its majestic divine realm to create a gigantic, red bubble-womb. Within the first few seconds and minutes of this rapturous explosion of love energy, all the particles (quarks, atoms, neutrinos) and packages of energy (photons) that now inhabit the universe are created. This begins the creative journey of the universe in space and time.

You will notice that I talk of Ruah as "she" or "her." In the Bible, the word *spirit* is, at times, neuter. In the gospels, the Holy Spirit is masculine. But there have been some studies of the word spirit, Wisdom-Sophia, and Spirit-Spiritus, in the Hebrew and Christian scriptures, that have led some scripture scholars to posit the possibility that the Holy Spirit is the feminine part of the Trinity. I have followed their lead in designating Ruah as feminine. In addition, I have been able to balance the genders of the Quadernity of Divine Beings in my story of creation; Sophia and Ruah are feminine, while Elohim and Dabar are masculine. (Notice that I will use Divine Beings, Godhead, Divine Friends, Deities, and other such names to designate this Quaternity of Persons.)

For the theological purist, I am not denying my belief in the Trinity of Three Persons in one God, but I am claiming my privilege as a writer to tell my tale in this midrashic manner.

I understand in using four persons, Elohim, Sophia, Dabar, and Ruah, in dialogue with the atom and photon characters in this story, I may be accused of "creating" four separate Gods. In speaking about the Trinity of Persons, theologian Elizabeth Johnson writes that R. Kendall Soulen has developed a framework to understand the oneness and the threeness of the Trinity.

> The thesis runs thus: the name of the holy Trinity is one name in three inflections. According to the dictionary, an inflection is a modulation of the voice in speaking or singing, a change in the pitch or tone of the voice ... one tone corresponds to the first "person," one to the second, and one to the third. Each *inflection* refers to the triune God as a whole but does so in a distinctive way ... (Johnson 2007 215–216).

Thus, when the second "person" emanates and distinctively inflects the whole of the one God, this distinct inflection does not take away from the wholeness of the other divine inflections, the first "person" and the third "person." When the first "person" distinctively inflects the wholeness of the one God, this inflection in no way diminishes the wholeness of the second "person" and the third "person." The same applies to the inflection of the third "person." When my midrachic story speaks of a Quaternity of Persons, I mean them to be distinctive inflections of the one God, which does not diminish the wholeness that each shares with the one God. When one "person" inflects, this does not diminish the wholeness of the other three "persons."

In spite of all our attempts to understand the one God through analogy and anthropomorphism, God is essentially incomprehensible, inexpressible, beyond description. Yet, as time brings on inexhaustible changes through the eons of our universe, our planet Earth, and its community of beings, especially us human beings, theologian Elizabeth Johnson intuits the need to press forward to find explanations of our origins, ultimate meanings, and final destiny.

> People keep on journeying through beauty and joy, through duty and commitment, through agonizing silence and pain, toward greater meaning and deeper union with the ineffable God to their last breath (ibid., 13).

This book is an attempt to press forward to discover "greater meaning and deeper union with the ineffable God."

During my daily walks around Saint Joseph and Saint Mary Lakes on the campus of the University of Notre Dame thirteen years ago, I mused on how to approach the writing of this story of creation. With cranial endorphins flowing profusely as I walked, a possible approach to my writing this story appeared mysteriously. *I should write the story as if I had just awakened from a vision of the 15 billion years of creation history depicted as exactly one year in length.* So, I, the author of this book, am the visionary and storyteller. A few days later, my mind was stimulated during another stroll around the lakes to depict the primal particles and atoms as characters to tell this creative tale in dialogue with the Divine Beings. (Particles and atoms, sometimes taking on new identities within various chemical compounds, continue to exist all through the 15 billion years of creation history.) As a visionary, I would possess special gifts of microscopic eyesight and instantaneous travel to any place in the universe. This would allow me to tell parts of the story of creation to supplement the above dialogue.

Having devised the setting and method for telling this creative tale, I was now faced with the decision of how to relate my atom characters to the Divine Beings. Would the personified atoms be directed in every minute detail to carry out a divine predetermined plan, or could they in some manner be creators? Auto poetic means that nature—photons, particles, and atoms, and later, the whole community of living beings—possesses the ability to decide and to direct, as creators, many of the events in the evolutionary process of the universe. More and more scientists who study the evolutionary story write about their belief that not only do biological beings possess consciousness, but photons and particles, like atoms and quarks, also possess consciousness on a certain level of their existence. In a true sense, they become creators. So this story of creation is written with the whole community of beings acting as creators.

Since my atom characters live in the microcosm of planet Earth (I would need a powerful electron microscope to view their creative activities), I needed to expand this creation story to help you, the reader, understand what was happening in the wider macrocosm of the Earth and universe. So I depict the Divine Beings taking the atom characters on frequent back-to-the-past journeys to visit other atom families creating other parts of planet Earth and the universe.

In writing this story from an auto-poetic point of view, I am not denying that the God is transcendent, the ground of all being, indescribable, and indefinable in human words and thoughts. I am inclusive in my theology. Paradoxically, I also believe God is eternally imminent and personally present to the universe and its community of beings. I believe deeply that the Godhead has been present accompanying and relating at a deep level the creation and evolution of the universe. Thus, the Godhead in whom I believe is not just a transcendent God, but an imminent, dialoguing God.

In keeping with my deep belief in the presence of the imminent God to the universe and its beings at a deep level, you will notice that Ruah is especially active at crucial moments in the creative process. .

I have written this account of the creation story for certain purposes. First, I wanted to express my faith in a Godhead of persons who are bound together by an inexpressible but profound, infinite mutual love and who have birthed forth in love the beginning of the evolutionary process of the universe, and later utilize the creative efforts of its community of beings. Second, I have striven to capture the reality of our universe as a web of creation intimately united through gravity, electromagnetism, weak and strong force. I believe that the unified source of these four energy forces is found in the infinite ecstatic love energy emanating from the Divine Beings. Third, I have attempted to show how, through the evolutionary process, each successive event of creativity has led to the next event. This process continues today. Thus, there is a 15-billion-year continuity existing in space and time between the first second of creation and the evolving existence of the present community of beings, including us humans on planet Earth and this entire universe. I want to highlight for you, members of the human race, presently beset with worldwide ethnic, racial, religious, and social conflict and violence, that our companions in the evolutionary process, the photons, quarks, atoms, neutrinos, and diverse biological beings, have much to teach us about the nature of self-sacrifice, compassion, and cooperation. If we are wise and humble enough to be open to this learning, they can teach us that we are really bound together with them as a family of beings living on spaceship Earth. We need to shed our arrogance, greed, selfishness, pride, and proneness to violence to join them in building a community of cooperation, compassion, and love on this blue, green, brown planet Earth.

Over and over again, as I speak to groups about the scientific story of creation, they have expressed an inability to grasp the element of time in the story. When I speak of 15 billion years of the evolving universe, or the fact that our sun, Earth, and solar system began about 4.6 billion years ago, they are unable to understand and grasp this kind of time. A few years ago, I was introduced to a method of bridging this seemingly insurmountable problem. This was accomplished by equating the 15 billion years of creation to the 365-day year. This has made evolutionary time much more understandable to the ordinary person.

In writing this scientific story of creation, I have had to make choices as to which theory to follow in relation to the formation of galaxies, the existence of dark matter, dark energy, black holes, neutrinos, and dates and times concerning the origin of planet Earth, biological life, etc. For example, I have chosen Lynn Margoulis's theory of the "symbiosis of cell evolution" over other theories. All the sciences are using their methodologies to collect data about nature to develop new theories. This is advancing steadily our understanding of nature, our planet Earth, our sun and solar system, and the whole universe. Often as I have read the newspapers and magazines or viewed television programs over the past twelve years, I have been confronted with new knowledge about our universe that has caused me to look at my writings to revise and bring them up-to-date.

I hope you enjoy the story of creation and the creative adventures of God, Quarkie, Photie, and their atom friends who, with their families of atoms, have created this majestic universe in all of its wonders!

Edward Ruetz

September 1, 2009

In Gratitude

I am grateful to Dr. Robert Ruetz, Dr. Edward J. Rielly, Bro. Lawrence Unfried, CSC, Dr. Sperry Darden, Bro. Raymond Papenfuss, CSC, and Dr. George Carter for their encouragement and advice to me on various portions of the book and for their critique of its contents. I am grateful to Sr. Mary Baird, PHJC, Sr. Janis Yaekel, ASC, and Sr. Nancy Raboin, PHJC, for their profound ecological vision that inspired me to begin and persevere in writing this book.

Characters

The names of the Quaternity of Divine Beings who begin the birthing of the universe and its community of beings are taken from the Hebrew and Christian Scriptures.

Elohim—From Genesis 1, 1f.

Sophia—From the book of Wisdom.

Dabar (Word or Creative Energy)—From Gospel John 1, 1f.

Ruah (Spirit, Wind)—From Genesis 1,1f.

These are the particles, quarks, atom, and photons that have been personified:

Quarkie—Female quark, later a part of a proton, part of helium 4 nuclei, and then a carbon atom, then hydrogen cyanide (HCN), part of the crew of Prokaryote Bacteria 1 or Prokye 1, which enlarged Prokye 1 and built nucleus membrane wall around DNA helix, part of crew of Eukaryote 1 cell or Eukarye cell 1.

Quarko—Male quark, later a part of a proton, part of helium 4 nuclei and then a carbon atom, then hydrogen cyanide (HCN), part of crew of Prokaryote Bacteria 1 or Prokye 1, which enlarged Prokye 1 and built nucleus membrane around DNA helix, part of crew of Eukaryote cell 1 or Eukarye cell 1.

Quarkoff—Male quark, later part of a proton, part of helium 4 nuclei, then a carbon atom, then hydrogen cyanide (HCN), part of crew of Prokaryote Bacteria 1 or Prokye 1, which enlarged Prokye 1 and built nucleus membrane wall around the DNA helix, part of crew of Eukaryote cell 1 or Eukarye cell 1.

Photie—Male photon, much later an ionized hydrogen atom, part of crew of Prokye 1, later part of crew of Prokye 3, which developed the mitochondria, then part of Prokye 1 again by symbiosis, then part of crew of Eukaryote Bacteria 1 cell or Eukarye cell 1.

Hydro—Male hydrogen atom, annihilated in a black hole.

Hydrojean—Female hydrogen atom, then part of an oxygen atom, then part of formaldehyde (CH_2O), part of crew of Prokaryote Bacteria 1 or Prokye 1, later part of crew part of crew of Prokye 2, which developed a tail assembly and spindle apparatus, came back to Prokye 1 by symbiosis, then part of crew of Eukaryote cell 1 or Eukarye cell 1.

Nitro—Male nitrogen atom, later part of ammonia (NH_3), then a part of Prokaryote

Bacteria 1 or Prokye 1, then part of crew of Prokye 2, which developed the tail assembly and spindle apparatus, returned to Prokye 1 by symbiosis, then part of the crew of Eukaryote cell 1 or Eukarye cell 1.

Phosie—Female phosphorus atom, then a part Prokaryote Bacteria 1 or Prokye 1, became part of crew of Prokye 4, which developed process of photosynthesis, returned to Prokye 1 by symbiosis, then part of crew of Eukaryote cell 1 or Eukarye cell 1.

Phoson—Male phosphorus atom, became part of Prokye 2, which developed tail assembly and spindle apparatus, came back to Prokye 1 by symbiosis, then part of crew of Eukaryote cell 1 or Eukarye cell 1.

Nitrojoan—Female nitrogen atom, became part of Prokye 3, which developed mitochondria, came back to Prokye 1 by symbiosis, then part of crew of Eukaryote cell 1 or Eukarye cell 1.

Oxydon—Male oxygen atom, became part of Prokye 3, which developed mitochondria, came back to Prokye 1 by symbiosis, then part of crew of Eukaryote cell 1 or Eukarye cell 1.

Oxyjoy—Female oxygen atom, became part of Prokye 4, which developed the process of photosynthesis, came back to Prokye 1 by symbiosis, then part of the crew of Eukaryote cell 1 or Eukarye cell 1.

Hydrojohn—Male hydrogen atom, became part of Prokye 4, which developed the process of photosynthesis, came back to Prokye 1 by symbiosis, then part of the crew of Eukaryote cell 1 or Eukarye cell 1.

These are the Prokaryote Bacteria and Eukaryote Bacteria characters:

Prokye 1—First prokaryote bacteria, developed enlarged size and a nucleus membrane wall around DNA helix, then lost its identity and existence.

Prokye 2—First replication of Prokye 1, developed tail assembly and spindle apparatus, later came back to Prokye 1 by symbiosis, then lost its identity and existence.

Prokye 3—Second replication of Prokye 2, developed mitochondria, later came back to Prokye 1 by symbiosis, then lost its identity and existence.

Prokye 4—Third replication of Prokye 3, developed process of photosynthesis, later came back to Prokye 1 by symbiosis, then lost its identity and existence.

Eukaryote Cell 1 or Eukarye Cell 1—The first eukaryote cell.

These are the colors of the antiparticles, particles, photons, and atoms:

antiparticles (black)

quarks (white)

protons (white)

neutrons (silver)

electrons (amber)

neutrinos (emerald)

photons (yellow)

positrons (topaz)

helium 4 nuclei (gold)

hydrogen atom (orange)

carbon atom (black)

oxygen atom (blue)

nitrogen atom (red)

phosphorus atom (saffron)

Abbreviations:

myich—Million years into creation history (from the beginning of the universe).

byich—Billion years into creation history (from the beginning of the universe).

bya—Billion years ago (from the twenty-first century).

mya—Million years ago (from the twenty-first century).

tya—Thousand years ago.

hya—Hundred years ago.

by—Billion years.

my—Million years.

ty—Thousand years.

ya—Years ago.

Chapter 1
Vision of the Quaternity

"Marvelous! Stupendous!" These words come out spontaneously. I am just awakening from an awesome vision! My mind is filled with vivid images—sights, sounds, and events—that I had witnessed concerning persons who were unimaginable and otherworldly. But I saw them!

Let me tell you what I saw and heard. The vision lasts exactly one year. And yet, it begins before all time. It comes from all ever.

I awaken one day in a land of brilliant light. My eyes are so overwhelmed by its otherworldly brightness that I am unable to recognize any familiar Earth marks. There is no brown or green land; no high mountains or broad valleys; no serene rivers or rushing streams; no blue sky and fleecy white clouds; no light and darkness; just blinding light. *Where is this place?* I think to myself. My eyes are almost blinded by the penetrating intensity of its white radiance. The scene is bathed in a broad symphonic movement of joy created by melodic strings. This music bursts forth in exuberant nobility and exotic grandeur.

Squinting, I can barely identify some faint figures amid the brilliance. There are a woman and a man, eternally young and appealingly beautiful. I can scarcely make out the image of another figure. It must be the couple's son. He is strong, handsome, and majestic in his bearing.

A crimson energy of love engulfs the woman and the man and their son from beyond them. This radiance glows crimson white. It seems like pure, vibrating love energy. This source of energy bends over the threesome to embrace them with love, gentleness, and peace. The three respond to the intimate touch of this personal crimson energy, which has the appearance of a ghostly fairy godmother. The joyous sound of the melodious strings enhances and permeates this awesome scene.

I strain to listen to their conversation. I wonder who they are. I soon learn their names.

The woman speaks first and makes a proposal for the man's consideration. "Elohim, we need to share our life and love with other beings!"

Elohim is puzzled by her words. He questions her proposal: "But there are no others here but the four of us, you and me, our son, Dabar, and our spirit of love, Ruah. We don't need other beings to make us happy."

Dabar understands Sophia. "You mean that we need to reflect outward our creative energy of love."

"Yes, this is what I mean. We share a profound ecstasy of love." Sophia's eyes brighten and I am almost blinded by their glow. "We must radiate this love that unites us. This outpouring will be our ecstasy of love. Let's send forth our ecstatic love and discover what can happen."

Ruah, the crimson spirit, who has been silent to this point, speaks up. "Sophia, you would send me into the blackness outside our radiant, loving sphere to allow this ecstatic love to create."

"Yes, yes, that's what I mean!" Sophia exclaims.

Elohim questions Sophia with a sly smile. "Our energy of love would then possibly create new beings! Isn't that interesting!"

I am aware that they all have embraced Sophia's idea "to create" and are ready to send Ruah into the darkness outside of their brilliant sphere. I also intuit that the energy of love that unites them through Ruah is unlimited. If Ruah is sent forth, they will in no way be diminished in their ecstatic love and unity.

Sophia intuits what the possible new beings will be. "When we send Ruah forth in our joint mission of bringing our creative energy beyond ourselves, we will create two new beings. We will call one of them time. Its first day will be called January first. The other new being will be called space."

"An excellent creation, Sophia! I embrace this new being, time." Elohim is quick to compliment Sophia's creativity. "Since we live in all ever, the eternal now, time will have a new character. It will possess a past, a present, and a future. Interesting! The eternal-now intimately present to past, present, and future time. Exciting!" He smiles as he concludes.

"Out of our timelessness we will create time and possibly other beings," Sophia says. "The possible other beings will live in our second creation, space. It will be interesting to discover in the fullness of time just what will happen to the other new beings who will live in the space we will create for them."

"Sophia, since space and time will exist simultaneously, we will call them by one name." Elohim suggests unifying these beings. "Forever, these entities will be named space-time."

"Excellent suggestion, dear Elohim!" Sophia affirms his naming.

"It's thrilling!" Dabar adds his affirmation to that of Sophia. "Space-time is well named." He pauses and then continues. "What will our creativity produce in space-time? How will these possible creations relate to us?"

"Maybe our creativity will bring forth beings to whom we can relate and share our love and unity in a special way." Sophia is looking to the culmination of their sending forth of Ruah.

"I am ready to go forth!" Ruah proclaims with enthusiasm. "Through our ecstatic love, we will create space-time as a home in which these possible new creations will live!"

Chapter 2
Creative Dawn

Standing outside of the brilliant radiance of the four, I observe the darkness. Indeed, it is very dark, like the dense blackness of a windowless closet. I wonder what the sending forth of Ruah in the act of birthing will do to its blackness. I am not ready for what really happens. I have not measured sufficiently the stupendous intensity of their ecstatic love energy.

"Go forth and birth new beings, Ruah!" Sophia commands.

"Yes, go forth and create new beings!" Elohim urges.

"We go with you on your journey, my dear Ruah!" Dabar cries.

What happens next is unthinkable. Space-time seems frozen in that first minute fraction of a second. I view the events that follow in very, very slow motion as micro-parts of a second of time. I experience each micro-part of a second as if it is a twenty-four-hour day.

Ruah deflects from the three into the darkness outside. In the beginning, I notice a tiny, solid black bubble, the size of a pea. In that diminutive ball, a new creation, ecstatic love energy, is packed solid. All of the energy and mass that will exist in future space-time is densely crammed into this very small object. Ruah's ecstatic energy of love, as a unified force, heats this pea-sized womb to 100 trillion trillion degrees. In a micro-part of a second, driven by this super-heated environment and enormous density, it begins to expand outward in all directions. In this wee part of the first second, this black, energetic pea expands and becomes a gigantic bubble. It takes on the red brilliance of trillions upon trillions of exploding hydrogen bombs. I view it as the most spectacular reddish sunrise I have ever experienced. My body glows red reflecting this awesome scene. Ruah hovers over this brilliance. The heat is overwhelming, like standing in the midst of millions of white-hot furnaces.

Within just the tiniest part of the first second, the ever-expanding bubble rolls back the blackness like waters suddenly being moved before a huge tsunami. Within this explosive energy of love violent collisions of gigantic numbers of massless yellow packages of energy begin to give birth to multitude upon multitude of very tiny strings and loops of string possessing mass. This energy and mass are impelled outward at tremendous speeds. (My visionary gift allows me to magnify the less-than-micro microscopic-sized packages of energy and strings and make them easily visible.) A turbulent snowstorm of these very tiny strings and loops fills the newborn space-time. Although the diameter of this bubble has expanded tremendously, the density of this newborn space-time is still extremely high, similar to the consistency of tar.

As I stand inside this crimson conflagration, I can hear the strains of a beautiful symphony. As I look closely, I notice that there are eighteen varieties of strings and loops of string. Each type of curved string vibrates and resonates. Like a violin string drawn tight to give a special sound quality, each oscillates and reverberates a unique, harmonic pattern. Their blended sound creates a magnificent musical masterpiece, a birthing hymn.

At this point, the harmonic strings and loops are uniting in different combinations. They give a musical birth to two distinct types of particles, white and black, originated by their colossal collisions with the yellow packages of energy. Space-time is like an enormously dense particle soup. In multitudes of crashes and interactions, the white ones become black, the black ones white. In this very, very brief moment of the first second, chaos reigns!

In the beginning of this drama, as I stood outside of Space-time, I thought my ears were deceiving me. I expected an extremely loud report, like multitudes of nuclear bombs exploding. But the sounds of this explosion of ecstatic love energy were like a rush of passion, overwhelmingly savage, yet melodically graceful, a musical undercurrent accompanying the dancing, darting particles that collided and careened into each other in this magnificent, expanding, brilliant red, womblike bubble. Now I realize this is the work of the melodious strings and loops. There is no doubt that Ruah is the catalyst in this symphonic outpouring amid the chaos of a birthing process.

I see Ruah's ghostly crimson-white figure hovering over this primitive chaos. Part of this intense love energy breaks off to form almost invisible energy lines connecting these quadrillions upon quadrillions of white and black particles and yellow packages of energy, which seem infinite in number.

Nearby Ruah, I see Elohim, Sophia, and Dabar. "What are those invisible lines of energy you are creating between the particles?" Dabar is inquisitive.

"I want all of these new beings to be connected intimately through our love energy!" Ruah gives her explanation. "I am calling these energy lines, gravity. It's a weak force, but it assures that all of our new beings will be forever united to each other and us in their new, expanding home."

"Through this weak force, gravity, you are joining these particles into a web of beings." Sophia says.

"Yes, a web of beings. No being coming into existence will ever be out of contact with all other beings," Ruah adds. She ponders briefly and continues: "But what name should be given to their new home?"

"We will call it the universe, since all are intimately united by gravity in the web of beings we have birthed!" Elohim says.

Chapter 3
Photie and Quarkie, the First Creations

Within a millisecond, the universe bubble-womb has grown to vast proportions. As the blackness is pushed back, its stunning expansion is accompanied by a musical harmony that begins to break up into clashing dissonance that overwhelms my ears. This climactic surge awakens in me expressions of hope, which stirs memories of the inspirational music of Jean Sibelius.

But there seems to be a remarkable change occurring. The expansion of the universe slows. Within a tiny part of this first second of its history, the temperature has cooled to 10 trillion degrees. Yet, in this red, glowing particle soup, the density of the fragments is still enormous, like the consistency of molasses. All of the particles in this dense, reddish soup are to be the raw material for all the future beings that will inhabit the universe. To the beat and harmonies of the strings and loops that make up the white and black particles, they continue to collide and interact with the yellow packages of energy. They move in all directions in this spatial chaos. Jean Sibelius would liken this phenomenon to the harmonic logic underlying a multiplicity of clashing dissonances. The music I hear ends with resounding hope for future deliverance from this turmoil. Soon, the black particles are annihilated and only the white particles exist. Why? Maybe like Sibelius's harmonic logic, underlying the musical chaos there is a law that deems that out of chaos comes order. I report only what I observe.

These microscopic white particles continue to collide with each other at speeds similar to a large asteroid colliding into our Earth. The weightless yellow packages of light energy are companions of the white particles. The former outnumber the white particles by quadrillions. Space-time is still very dense.

I observe Ruah brooding over this brilliant, reddish-colored uterine bubble, the universe. Nearby, Dabar, Sophia, and Elohim glow radiantly.

Dabar speaks first. "We need to establish some rules that will govern these new beings, the white particles and the yellow packages of energy."

"You are correct in your judgment, Dabar." Elohim has been thinking along the same line. "Our love energy that birthed the universe can change from one form to another and back again to its original form. It can never be destroyed and new energy made. This is the first law."

"So what you are saying is that the total amount of energy in the universe will always remain constant." Sophia summarizes succinctly the law that Elohim has proclaimed. "We can call this the law of the conservation of energy."

Elohim has an additional law to propose. "The second law is this. It will possess all the love energy it will ever have in its history. With every energy exchange from one form to another, disorder will increase in the universe. So disorder is more likely than order to occur through space-time."

Sophia, who had summarized Elohim's first law, now proposes a law. "We will decree that the number of the weightless yellow packages of energy that make up the total amount of energy will outnumber the white particles by quadrillions upon quadrillions."

"Three very important laws!" Ruah agrees with the decisions of Dabar, Elohim, and Sophia. "From now on, within this dense, superheated particle soup, the new beings created will obey these three simple laws."

"Let's talk to these new beings," Elohim says, "and through our conversation we can assign them names."

"The yellow particle is not only full of energy, but also light." Sophia is looking intently at the first yellow being created. "Since you are yellow in color and our first creation, we will call you Photon 1." Sophie approaches this new being. It is traveling at tremendous speed. "Greetings, oh beloved Photon 1, so full of energy and yellow light. You and your vast family will be a source of light for the universe. We have birthed this red bubble in which you exist! I am Sophia. Let me introduce you to Elohim, my husband, Dabar, our son, and Ruah, the personal love that binds us together in unity and love." With the introductions complete, Sophia, her maternal being full of sensitivity and nurture, finishes with her final remarks. "But let us be more personal and call you Photie. We love you, oh dear one, and invite and empower you to be a creator."

Still stunned by his new existence and the dense chaos surrounding him, Photie listens intently to Sophia's words of naming and empowerment. Moving at the breakneck speed (I recall from my high school physics class that light travels at 186,282 miles per second, no more, no less), Photie, filled with deep anxiety and lacking a sense of personal identity, makes no response to Sophia's greeting but speeds off into the almost limitless, but expanding, space-time as a speck of light. "Photie did not stop to exchange greetings, did he?" Dabar remarks as Photie speeds off into the distant reaches of space-time.

"He is extremely frightened by his chaotic environment and is stunned by his sudden birth into existence and space-time." Sophia empathizes with Photie. She worries about his present state of fear and how he will cope with his new life.   "Sophia, possibly we will have a chance to meet him in the future." Dabar then turns his attention to the white particles. "They shall be called quarks!" He is emphatic.

"I like that name, quark," Sophia affirms. "It has a very distinctive ring to it. Let us greet them!"

Ruah gazes at these white particles and spots one in particular and says to the Divine Trio, "Let me introduce you to our new being, Quark One."

Sophia is the first to greet Quark One. The former observes that she is full of fear and feeling desperately lonely. "How are you, Quark One, my dear one?" Sophia pauses and continues. "Allow us to call you Quarkie, for short."

Quarkie draws into herself. She is very shy. She summons the courage to haltingly respond to Sophia. "Who … who … are … are … you, oh … ra … ra … diant one?"

"I am Sophia. Let me introduce you to my husband, Elohim, our son, Dabar, and Ruah." Sophia observes that Quarkie is extremely befuddled. She adds, "We gave birth to this red bubble home where you live! But don't be afraid. We wanted to share our deep love and happiness with other beings. We sent Ruah to roll back the blackness around us and birth this home for you, our new beings, with whom we can share our happiness and love. You, Quarkie, with our departed friend Photie, are the first beings we have greeted. How lovely you seem, my dear!" After a pause, she smiles and exclaims, "Quarkie, you are formed out of small, curved musical strings and loops of strings. You vibrate with melodious harmonies."

With care, Quarkie ponders the tale that Sophia has just spun. But she discovers more than the story of her origins in Sophia's words and attitude; namely, that Sophia loves her very deeply. Further, she is beginning to trust Sophia, Elohim, Dabar, and Ruah. Yet, she needs to know more about them and herself. Fear is fading, but she is filled with loneliness. "I ... I ... feel so very lonely in this new existence." Gaining courage, she says, "But what am I to do? Keep moving farther and farther out in space-time like that yellow package of energy? What is my purpose for being here?"

Dabar seems sensitive to her feeling and her quandary. Yet, he gives a strange answer to her questions. "That is for you to decide, my beloved friend!"

This very, very tiny particle, Quark One, looks deeply perplexed. "I do not know what to do. I look to you for guidance, my dear friends!" With expectation, she awaits their answer.

Elohim is the first to respond. "Quarkie, what do you want to do right now?"

She waits and ponders for a short period of time. Finally, Quark One says, "I need companionship. I need to get to know others of these new beings who are like me. But ... but I am afraid to make the first contact."

Ruah speaks. "Quarkie, I have connected you with all the other particles in this vast space-time where you live. Through a weak love energy, gravity, you are bonded to all these other beings. You are not only connected, but gravity attracts you to each other. You were made for each other, believe me."

"Don't be frightened, dear one!" Sophia tries to reassure her.

This response leaves Quarkie with mixed feelings. She is afraid. Yet she longs for friendships.

"Quarkie, what are you going to do?" Dabar prods her with gentleness.

Quarkie ponders some more. She draws a deep breath. She muses some more. Finally, she breaks the silence. "I will try to make friends with these other beings."

"Quarkie, follow your deep-felt need for friendship! We love you. We believe in you!" Dabar encourages her.

When Ruah breathes upon her in love, Quarkie musters additional courage and strength. "I will make friends!" she proclaims. She looks around and sees some nearby white particles. She feels an attraction to two handsome quark particles. Will they form a loving trio of friendship? With grace and courage, she says simply, "Oh neighbors, I would like you to be my friend. My name is Quarkie."

Her neighbors feel the same attraction. They smile and welcome Quarkie with sincere openness. They embrace. The new friend says, "I have felt so fearful and lonely in my very short existence. I too have been longing for a friend. Quarkie, you are my friend."

"You need a name, friend," Elohim says. "Your name is Quarko."

The third quark finally speaks up. "Quarkie and Quarko, I am so, so glad to join you in friendship. I also feel so alone."

"Let me name you, dear quark." Sophia seems to relish this act of naming. "Your name will be Quarkoff."

They meld together in love and friendship. Their union is very important for the next step in the growing process of the universe. I observe some of the ecstatic love energy driving the expansion of the universe diverted toward this trio of quarks binding them tightly together.

Dabar commends them for their courage in joining into a community. He gives them some additional information. "You remember, Quarkie, that Ruah told you that you were bound together with all other particles and energy packages in your new home, the universe, by a special power called gravity. And so you are. Now I inform you, Quarkie, Quarko, and Quarkoff, that the three of you are bound together by an additional power called a strong force. The latter's power is derived from our energy of love. All trios of quarks are bound by that same power."

In awe, I watch them move out together into the open, reddish-glowing space-time before them. The message of their friendship spreads rapidly through the quark community. Soon, more and more quarks are establishing trios of friendship through the dynamic energy of the strong force emanating from Ruah. A loving community is being built within the first milliseconds of creation.

"These communities of quarks need a name!" Elohim is musing as he asks, "What shall we call this new being?"

Sophia is quick to respond. "It is our first permanent being. We shall call it a proton."

"First it is!" Dabar exclaims with excitement.

"Sophia, a wise choice. A proton it shall be called." Elohim agrees. "Protons will be white and filled with positive energy."

"Quarkie, Quarko, and Quarkoff, you are creators! You have created Proton One. Your many trios of quark friends are creators too!" Ruah is filled with joy. "Go forth into expanding space-time with the multitude upon multitudes of the energy packages, the yellow photons that were birthed in your early spatial collisions."

"I wonder what our new beings, the quarks, now protons, will create through their community of friendships," Elohim muses.

"We have just begun a process, the empowering of these new beings to be creators," Sophia adds. "Where will their creative powers lead them?"

"I can't wait to find out," Dabar comments.

I see the crimson radiant one, Ruah, breathing her energy of creative love on the edges of this expanding, brilliant, reddish space-time. I too wonder, how will this *story end?* It is still the first thousandth of a second very, very early on January first!

Chapter 4
The Birthing of Neutrons, Electrons, and Neutrinos

It is within the first second after the birthing of the universe. The temperature within the rapidly expanding spatial bubble is still raging at about 10 trillion degrees. Time is still running in very, very slow motion. The next few moments within this first second again seem to me like a day in length.

As the womb of the universe grows enormously in diameter, I observe new particles being born out of the collisions of two yellow photons. At this extreme temperature and high density, a Star Wars battle of epic proportion is raging to the beat of lyric strains embellished in Chopinesque tone-colors. Photon strikes photon at such enormous velocities that they give rise to new microscopic particles: an amber particle and a topaz antiparticle. Energy is transformed into mass.

From a blanket of silence, the soft sound of a Chopin theme arises amid ominous signs of the onset of a ferocious storm. To my left, the discordant sounds of individual struggles as an amber particle crashes into a topaz particle. To my right, a rapid musical passage builds in ferocity to a climactic resolution as particles annihilate each other, freeing their energy to create two massless yellow packages of energy, photons. Mass is converted into energy. The new rules also determine the basic forms that mass and energy will take. By obeying these laws, the yellow packages of energy and the white, amber, and topaz particles are in dynamic equilibrium, so their respective numbers are maintained. These events are happening all over the ever-expanding, crimson space-time to the accompaniment of the lilting harmonies of the curved strings and loops living inside of the white, amber, and topaz particles. This signals the end of the musical epic and of these individual battles.

The various particles and yellow packages of energy are part of the less dense particle soup moving through the expanding universe. Amidst the hellish heat of that first second, an amber particle flying through space-time collides violently into a beautiful white proton, giving birth to two new particles, one silver and one emerald. The silver particle, with similar mass to a proton, joins the particle population. The energized emerald particle zooms off into the rich, reddish womb of the universe. These collisions are happening all over the growing universe. Now space-time is populated by vast numbers of white, amber, silver, topaz, and emerald particles and yellow packages of energy.

Quarko, Quarkoff, and Quarkie, as Proton One, are involved in a violent collision with an amber particle. They are aware that a new particle is being created. They begin to discuss their experience.

"I can feel that we have given up something of ourselves to bring into being these silver and emerald particles, but paradoxically, I don't feel we have lost anything of ourselves." Quarko is full of wonder as he speaks to his mates.

"Bravo, Quarkie, Quarko, and Quarkoff!" Ruah hovers over them and initiates a dialogue. "You are creators with us. You are our beloved beings."

Quarkoff is listening intently to the conversation. He comments, "I could sense your presence, Ruah, and the intensity of your love energy when the amber particle struck us. This love force helped us to create these new beings."

Quarkie's spirit intensifies through this experience. "Thank you, Ruah!" she says. "It was the energy of love with which you have gifted us that helped us to birth these new beings." She has a question to pose to Ruah. "What names will you, oh Divine Beings, give to these four new particles, the amber, topaz, silver, and emerald ones?"

Unbeknown to me, Elohim is nearby and begins to address this question. "The silver particle has about the same weight as you, oh beloved trio, but it has a neutral charge. We will call it a neutron, your intimate partner in the universe."

"An appropriate name, my beloved!" Sophia is quick to affirm Elohim's choice of a name. "Let me give a name to the amber particle." She ponders for a moment. "We will call her an electron. The 'electr(ic)' part of her name means generating a negative charge. The 'on' means to be in constant motion. Thus, the amber electron is a lightweight particle, negatively charged, my dear Quarkie, Quarko, and Quarkoff. Between you and an electron, there is a huge difference in weight. The amber electron is many, many times lighter than you, my friends, and your companions, the silver neutrons."

"Could I name the emerald particle?" Dabar asks. He goes on. "It bears no charge like the neutron, is practically without weight, and moves almost as swiftly as the yellow photon. It is like a tiny neutron. We will call it a neutrino. This emerald particle is traveling through space-time and seems to be able to move through any objects in its path."

Ruah ends the conversation. "Well named, Dabar, for the role it will play in the history of this unfolding universe." She pauses for a moment in deep contemplation. "We have not named the topaz particle. It is related to the negatively charged amber electron. Light in weight like the electron, it possesses a positive charge. We will call it a positron, the electron's antiparticle. It will play a part in the building of the universe."

The conversation yields to a period of contemplation. Finally, Ruah breaks the silence. "We are less than a second into the birthing of this new creation, the universe. Amazing! Where will this creative process of birthing end?" The crimson-white godmotherly figure glows intensely with fiery love that energizes all the quadrillion upon quadrillion of multicolored particles interacting dynamically in this particle soup of glowing, red space-time, still united by gravity.

I stand observing the awesome, unbelievable scene in deep wonder. I too am enveloped in contemplation. At the end of a long period of quiet, my mind proposes this question: *where will this story end?*

Chapter 5
The Red, Glowing Universe

One second old! The universe is one second old! Through my marvelous visionary gift, I find myself situated within this gigantic, crimson, glowing bubble-womb of the universe. It is now about two hundred thousand miles in diameter. (From my memory before this vision, I recall that this is about the distance from our Earth to the moon.) The bubble's temperature has cooled to tens of billions from the tens of trillions of degrees just a fraction of a second ago. As a result, the rate of its expansion has slowed significantly from its immense enlargement a few micro-parts of a second ago. Its color has a diminishing reddish glow, but is still very bright. Due to my amazing vision of the birth of the universe and its aftermath, totally and unexpectedly gifted to me, I am allowed to travel instantaneously to any part of the universe that I desire. I am able to magnify the micro-microscopic particles so that I am able to observe them in their environment and in their relationships with each other. I am able see the Divine Beings in all of their magnificent splendor and radiance. This visionary power allows me to give a personal account of all of the major events that are occurring in the history of the universe.

The particle soup within the bubble is still dense and full of violent activity, endless collisions, particles going in and out of existence. This dense stew is made up of lightweight particles, the amber electrons; the topaz positrons and emerald neutrinos; and the heavyweight particles, the white protons and silver neutrons. The massless yellow packages of energy, the photons, are the other beings in this bubble-womb. The photons, which vastly outnumber

20

the particles, are like massless shooting stars, specks of yellow light, traveling at unbelievable speed.

In the misty, reddish chaos of the universe, I observe a very mysterious dark matter and dark energy. It seems to be much more plentiful than the other more colorful particles. Dark matter and dark energy are ominous and characterless. I am unable to describe it. I ponder its meaning in this new universe. Maybe I will discover an answer for its existence and purpose in the future. This mystery within the red growing, glowing universe overwhelms me. I am in awe of its wonders.

As time moves forward, from one second to ten seconds to fifteen seconds, the expansion of the spacious red bubble slows down further. The particles' billiard ball game continues as the temperature cools to about 3 billion degrees.

As time approaches its second minute on early January 1, the creative power of this very new universe seems to have ceased. Within the third minute, Ruah is overarching this scene, radiating her crimson energy of love. At descending temperatures of about 900 million degrees, the attraction between white protons and silver neutrons is producing a new gold particle.

Ruah notes its birth to Dabar. "I wonder if Quarkie has become a part of this new creation." She wants to enter into dialogue with Quarkie, but she is nowhere in sight. "Two white protons and two silver neutrons have joined together to create this new gold particle. She could be part of it."

Dabar decides to greet the next neophyte particles passing by. "Oh ho!" he begins the conversation. "Have you seen one of your community members, Quarkie?"

This new gold particle shouts back. "Quarkie is right behind me in another gold particle."

Ruah is the first to greet Quarkie. "My dear friend, Quarkie, you are part of another new creation."

"Yes, oh Divine One, Quarko, Quarkoff, and I, a positively charged white proton, were strongly attracted to a neutrally charged silver neutron." Quarkie responds with enthusiasm.

"We positively jolted our new partner when we bonded," Quarko utters with a devious grin.

Quarkie disregards his smart remark. She goes on with enthusiasm. "But no sooner had we joined in this alliance when we encountered another white proton/silver neutron. We held a brief meeting. We were unanimous in our decision. 'Let's join in community with this new friend, the white proton/silver neutron.'" She hesitates. Then she continues. "We have birthed a new creation." Quarkie proudly proclaims their inventiveness.

Ruah addresses her. "Quarkie, in your union, it was the strong force of primal love energy that bonded you and your quark mates, the white proton, to the silver neutron." She pauses for a moment and then continues. "This strong force, like gravity, has its origin in a single energy source, our ecstatic love energy that gave birth to this expanding universe. The strong force is the glue that binds all of you together in this golden particle."

Quarkie is more interested in her new identity. "But what can we call ourselves now? I'm still a quark and part of a proton, along with Quarko and Quarkoff, but also a part of this new being."

"I will name this new being, Helium 4!" Elohim enters the conversation with gusto.

"Helium 4!" Initially, Quarko savors the new name. Then, with sarcasm, he adds. "That's a hilarious name—a heliotropic name!"

"You may be very prophetic, my dear Quarko!" Elohim commends his use of this adjective. "But you will have to wait to discover whether your prophetic name for Helium 4, heliotropic, will come true."

"I am a quotable, quasi-prophetic quark of deep profundity, Elohim." Quarko deadpans his humorous utterance.

"Well, we are graced by the presence of a homey humorist, my friends." A faint smile illumines Elohim's radiant countenance as he counters clever Quarko.

Quarkie has had enough of Quarko's devious humor. She abruptly changes the subject. "Does the Helium 4 have a meaning like proton, the first?"

"Two protons and two neutrons equal four." Sophia heralds a partial explanation.

"Yes, my dear Sophia. That explains the four." Elohim is ready to add his own explanation of the other word, helium. "I have invented this word because I have a premonition that this new element is destined to perform some important work in the future of the universe. We must wait patiently to find out what it will be and discover if our dear Quarko is a crazy or true prophet."

Quarko takes note of Elohim's ambiguous characterization of him but decides to remain silent.

As I stand in awe of this diminishing red sunset bubble, I observe silver neutron particles being attracted by the strong force of the energy of love to the white proton particles to form this new gold particle.

It is the fourth minute of creation history! As I look at the chaotic, colliding particle population, I notice that an amber electron begins to mate with the still single white protons. The bonding begins between them. Almost at the same moment, a speeding yellow photon strikes the amber electron and breaks this bond. These events are happening all over the universe. The temperature is much too high to sustain these electron/proton matings.

At the fifteen-minute mark of the universe's birthing early on January 1, the temperature is a cool 300 million degrees. The deep reddish-glowing universe bubble continues its gigantic swelling but at a still declining pace from its enormous early expansion fourteen minutes ago. The particles continue their chaotic activity. The density of the elemental particles lessens greatly as space-time expands. From the consistency of molasses about fifteen minutes ago, the density now is more like that of water. Creativity seems to be at a standstill!

It is twenty, twenty-five, forty minutes into creative history. Finally, the first hour of the first day of space-time arrives. As the universe grows, the density lessens, the temperature cools, and the deep reddish glow diminishes like a receding sunset.

I listen to a conversation between the Divine Beings.

"Since our birthing of the universe, notice the intricate web of relationships woven by gravity between all of its new beings," Ruah comments to Sophia. "Our ecstatic love energy is the source of this attraction between all of the new creations. The strong force of our energy of love attracts and bonds the white protons and silver neutrons creating helium 4."

"And all of our new beings are creators," Sophia responds to Ruah. "And thus it is unknown exactly how they will interact with themselves in the future." Sophia hesitates and then continues. "Let's wait patiently to see what will happen."

"I can't wait to find out how all of this will transpire." Dabar is excited as he speaks. "Yes, I am expectant and full of wonder!"

Chapter 6
The Creation of the Hydrogen Atom

Time marches onward. It is about three o'clock on the morning of January 1. The temperature is drastically moderated to about three thousand degrees. The universe continues to expand. The density of particles is lessening. Again, my visionary gift allows me to see clearly the immense, uncountable number of elemental particles of matter, which are submicroscopic in size. This newly formed environment is filled with a kaleidoscopic snowstorm of flakelike particles: white, topaz, amber, silver, gold, and yellow, all bonded through the energetic power of gravity. The late sunset-red glow of the earlier bubble-womb is almost extinguished. Presently, it is like the last bit of pink glow on the horizon before the coming of the blackness of night. Through these first three hours, the yellow photons, like lightning-fast spaceships, have continued to speed through space-time as flicks of tiny light. Over and over again, they impact the amber electrons just as they begin to mate with the white protons, preventing them from bonding. But a change is taking place. The yellow photons are losing their previous ability to prevent the particle population from mating.

Suddenly, I notice that more and more independent but negatively charged amber electrons are bonding with the positively charged white protons. The very swift yellow photons continue to bombard these newborn unions. Now, at this very moderate temperature, the yellow packages of energy just bounce harmlessly off the amber electrons. Previously, the high-speed, energized

25

photons under very intense temperature could strike the electron-proton bond, breaking them apart and preventing their union. Presently, they can tightly bond themselves together with the amber electron orbiting around the white proton nucleus. This process is happening all over the ballooning universe. Quadrillions upon quadrillions of new beings come into existence. They far outnumber the golden helium 4 nuclei by about four to one. Ruah, as a ghostly godmother, glows a bright crimson as she overarches the entire horizon of the darkening universe. Again, the primal energy of love which birthed the universe enables the white protons to join the amber electrons. This new being is orange in color.

Ruah comments to Dabar, "We are entering a new age in the lifetime of particles. Do you see this new orange being? It is being held together by a special energy force. Let us call it an electromagnetic force. Like gravity and the strong force, the source of its energy is our ecstatic energy of love. It keeps the amber negatively charged electron in an orbit circling the white positive proton and creates a new being."

"Since it is a new creation, Ruah, this being needs a name." Dabar is deep in thought. There is a pause.

Sophia eavesdrops on their conversation. She breaks the silence. "I have a premonition that this neophyte is destined for several important missions."

"You already know what to call this new being." Elohim intuits that Sophia is prophesying about its future and her readiness to give this new being a name. "Hydrogen is its name!" she solemnly proclaims. "The *gen* means that which gives birth. We will discover at some future time the meaning of *hydro*." She continues. "We will also call it an atom. Since Helium 4 has no accompanying electrons, it will be called a nuclei. I intuit that Hydrogen, with Helium 4 nuclei, will become the parents of immense families of other atoms."

Elohim comments softly, "Quite a prophecy, Sophia!"

Dabar greets a nearby hydrogen atom. "My dear Hydrogen One, welcome to the universe."

Hydrogen One appears puzzled and befuddled. "Are ... are ... you ... spe ... spe ... speaking to me, oh bright one?" Hydrogen One is feeling awkward and fearful. "Who am I? Who are you?"

"You are a new creation in this vast, darkening universe, which will be your new home." Dabar is orienting him to his new surroundings. "Members of your particle community, the amber electron and white proton, have created you. You are also destined to be a creator in fashioning this expanding universe. You are an atom, but a special kind of atom called hydrogen. That's a long name." Dabar pauses to ponder briefly. He then continues. "How would it be if we shortened it to Hydro?"

"I like my new name, Hydro." He is losing some of his shyness. He begins to warm up to the Divine Beings and his new home. He has a crucial question to ask. "I am here in the universe with you, but what am I to do now?" Hydro is serious.

"Look around you, Hydro. What do you see in this universe?" Ruah waits for his response.

"I see limitless black space. I observe myriad upon myriad of yellow dots of light moving silently and swiftly all over my new home. I see multitudes of orange hydrogen atoms like myself moving much more slowly in this space, along with the golden particles." I think he has ended his accounting, when he adds, "In fact, I am being attracted toward an immense number of members of my hydrogen family and the helium 4 family, who are lumping themselves together in various parts of the universe."

"Maybe you are answering your own question, Hydro." Ruah is affirming the attraction he feels to a particular lumping of hydrogen atoms and helium 4 nuclei in space-time.

"I can sense that my neighboring hydrogen atoms are feeling this same longing." Hydro pauses momentarily. "You told me that I am to be a creator. I like this mission you have given me." Hydro is quickly growing in confidence. There is a note of power and pride in his voice as he says this. He goes on. "I am going to follow my attraction to my new family. I feel especially drawn to that lumping of hydrogen atoms ahead of me! Possibly I can become a leader in this creative mission." Off into space-time he travels.

"Good-bye, dear friend Hydro!" Elohim and Sophia shout in unison.

Dabar muses deeply about the future. "Hydro and his friends, the hydrogen atoms, are destined to be parents giving birth to *what* in the future?" His thoughts go forward following their unique creation, space-time.

"I have a feeling we are very, very early in the journey with our new beings, the universe and space-time." Sophia is excited and expectant. Then she pauses. There is a worried expression on her radiant countenance that overtakes her exaltation. "Did you observe Hydro just before he took off?"

"Yes, there is some of the enthusiasm that Quarkie exhibited when she bonded to Quarko and Quarkoff." Elohim measures his next comment carefully. "Hydro seemed excessively proud about his empowerment as a creator. He appears to be savoring his personal power."

Sophia confirms Elohim's insight. "I perceived his grasping for this power. He is enamored by it." She pauses momentarily in deep thought. But a smile gradually begins to light up her already radiant face. "Yet, I am enthusiastic about the universe and its new beings." She looks at Ruah. With expectation, she asks, "Where will this journey of the universe take us next?"

I watch Ruah. She has a broad smile on her radiant, ghostly face. She is accompanying the new beings in their creative journey.

Ruah is hopeful about the future creativity. She says, "Do you notice, Sophia, that the hydrogen and helium 4 nuclei are gathering and lumping themselves together in the form of islands in space-time?"

"There are multitudes upon multitudes of these islands being formed by these new beings." Sophia observes the same phenomenon. "Just a few hours ago, all of these beings were in a chaotic state. And now they are beginning to bring some order and pattern to their universe home." She muses briefly. Then she adds, "I wonder why this is happening."

"Could it be that when there is great chaos, the next stage is that of order.?" Ruah comments. "Hydro and his hydrogen family, and the family of the helium 4 nuclei, made the decision to lump themselves together in these islands of gases. This must be part of their creativity."

"Ruah, let us wait patiently for the next step on this fantastic space odyssey. We have a long, long creative path to follow! Let us go forward with expectancy." Sophia concludes this conversation.

Chapter 7
A New Cocreative Mission
Is Begun

All over the blackening universe, I can see multitudes upon multitudes of the neophyte gaseous particles, orange hydrogen atoms and golden helium 4 nuclei, being attracted to billions of fast-growing, lumpy islands. These expanding islands of particle lumpiness are separated by gigantic desertlike stretches of space-time. From early on January 1 to late January 2, the atoms and nuclei are being drawn on a curved path like families of hornets to these billions of island nests.

Within each island, they are gathering into gigantic, gaseous clouds. As their density grows within different areas of these clouds, drawn by gravity's energy of love, hydrogen and helium 4 gases begin to rotate, creating enormous golden orange gas-filled spheres. The dark matter and dark energy hovering high above these islands are exerting intense pressure upon the elements in these gaseous bodies.

The giant rotating saffron gold spheres, driven by all these forces, cause the heavier helium 4 nuclei to fall into their center cores. Through a three-step process, hydrogen atoms near their cores are losing their electrons. They are becoming nuclei consisting of one proton. Gradually in the triple-step process, with the help of the weak force, they are being transformed into helium 4 nuclei to fuel the center cores of these spherical bodies. As this process continues, the temperature in their cores and surrounding area is intensifying to millions of degrees. The centers of these spheres are now raging infernos.

As I look back on all of these billions of lumpy, gaseous islands of hydrogen and helium, this same process is happening all over the expanding universe. In all of the multitudes of clouds within each gas-filled island, helium 4 nuclei, surrounded by the lighter hydrogen gases, are gathering in the cores of numberless spheres of spinning gases, some small, others medium, and some extremely large. These rotating, circular bodies of gas look like billions of Ferris wheels spinning on their sides at energetic speed.

Within one of these gaseous islands and in a particular agitated cloud, I observe Ruah and Dabar entering into the midst of one of these enormously large spinning infernos looking for Quarkie and her mates. Finally, they spot them.

Dabar greets her. "Quarkie, how are you doing?"

Quarkie is glad to meet them again. She responds, "I feel so happy working with my nuclei colleagues, Dabar. We are engaged in a great mission." She is enthusiastic. "I feel we are close to some new momentous event. It will be significant for this whole universe."

Dabar picks up her excitement. "You are enthused, Quarkie. It is so apparent as you greet us." He continues, "What do you think will happen?"

"I only know that our whole family of nuclei is totally involved with each other. There is a mutual respect and growing love between us." Quarkie sighs in joy. "We believe that we will have to give up our present existence for something altogether different."

Ruah enters the dialogue. "Doesn't change make you fearful and apprehensive, Quarkie?" With eagerness, she awaits Quarkie's response.

"Not at all, Ruah." Quarkie pauses to recollect herself. Then she continues. "I am ready to give up my present existence as a proton inside this helium 4 nuclei for the sake of moving the universe into a new age."

"You are deeply committed to our mission, Quarkie." Ruah acknowledges her resolve. "You have grown in courage and openness since we had our first encounter with you within the first second of the birthing of the universe. You are not afraid of change." Ruah has one more comment to make. "Because you are not afraid of change, Quarkie, you are an excellent creator. Change will probably always be an important process in your creative journey together."

"We also are not afraid of change, Ruah!" Quarko and Quarkoff shout in unison. "The whole community of nuclei is united in love and resolve," Quarkoff concludes proudly.

Dabar inquires, "What will you become if you lose your present existence?"

Quarkie gives a rapid response. "We feel confident in the future. We will wait patiently to see what will happen."

Ruah concludes the conversation. "We too wait expectantly to see what will happen, my friends!"

When the quark-mates depart, Dabar speaks to Ruah. "I notice that the mysterious dark matter about which you spoke earlier is dwelling over this island and involved in the quarks' mission. In fact, dark matter and dark energy reside like halos over all of the billions of gas-filled islands."

"Yes, Dabar," Ruah responds. "Not only dark matter and dark energy, but also the emanations of our ecstatic love energy, gravity, weak force, strong force, and electromagnetism, are intimately involved in the process that is going on in this rotating sphere of hydrogen and helium gases." Ruah directs Dabar to turn his gaze toward the universe beyond their location. "Look out toward all of the billions upon billions of similar gaseous rotations on the billions of lumpy islands, Dabar. The same process is going on there also."

* * *

Above the roaring, white-hot core of the helium nuclei in this same enormous rotating sphere, the hydrogen atoms are losing their electrons and are being drawn more and more closely together to carry out their transforming process. The surrounding orange hydrogen atoms become nuclei by shedding their amber electron partner. As nuclei, they are exerting great pressure on the golden nuclei core in the process of creating new helium 4 nuclei to fuel the core and raising the temperature of both of them.

Elohim and Sophia are present in this area of this gyrating body. They locate Hydro at a distance from the raging core and greet him.

Sophia begins a conversation. "My friend, we haven't talked for a long time. How are you doing?"

"Our hydrogen family is working together on a great project." Hydro is proud of their effort. "I sense a pressure thrusting me and my hydrogen family members deeper toward the core of this sphere. I have been trying to lead the effort to carry out our mission. I am encountering some resistance. But I believe our mission will end in something new."

Elohim is quick to ask, "Something new? What do you mean?"

"You told me when I was created as a hydrogen atom that the *gen* in my name meant giving birth to something." Hydro recalls their previous conversation. "I intuit that by my leading the effort, we will give birth to something entirely new to decorate the universe."

"I believe you, Hydro. I will wait with eagerness for this birthing event." Elohim supports Hydro in his premonition.

"I wonder what will be birthed." Sophia ponders her question and then adds, "Go forth to your creative birthing, Hydro. But don't rely on your leadership alone to accomplish it. Remember that you, the hydrogen family, are creators with the helium family. Ruah is the accompanist in this mission of furthering the creative process. Place your reliance on her for emotional support, my dear Hydro." Sophia is cognizant again of Hydro's spirit of independence and pride.

Hydro does not respond. He turns aside and is upset with Sophia for questioning his fidelity to the mission. He exits in anger.

"Hydro's prideful and independent nature is driving him," Elohim comments.

"I am fearful, my dear Elohim, that he will be the source of future trouble." Ruah seems deeply concerned about his attitude and its future consequences. Their conversation concludes.

To what will all of this activity of the hydrogen and helium families give birth? As I await their new creation, I stand in awe of this remarkable scene.

Chapter 8
The Nuclear Furnace Ignites

In the core of this gaseous cloud, orange hydrogen atoms and nuclei, including Hydro, work through the three stages of a process, activated by the four energy forces, to form new golden helium 4 nuclei. Two protons and two neutrons of heavy hydrogen, devoid of their electrons, are adding new helium 4 nuclei fuel to its raging hot core.

Suddenly, I am aware of a quickened, melodic sound. The subatomic strings and loops of strings that make up these elements are producing their creative musical vibrations to accompany the furious activity. *What is its meaning?* I ask myself. It has the sound of a rapidly moving presto pattern of one of Beethoven's classic Symphonies. The critical temperature at the core of this rotating sphere has reached 18 million degrees. Now I see this furnace ignite and a bright yellow mushroom fire begin to engulf the core. It is the momentous time of the first nuclear fusion and ignition in the short history of the universe. The bright, golden yellow flames of the ferociously burning helium gradually begin to send outward a multitudinous army of neutrinos and photons. The latter are arrayed in brilliant yellow light and spectral radiation, red, orange, yellow, green, blue, indigo, and violet, as they race through this orange hydrogen-saturated sphere toward its surface. This cloud has spun out its first gigantic yellow sphere, which illuminates its neighborhood.

Suddenly, the black bubble-womb of the universe observes the photonic light and multicolored spectral radiation of the first nuclear fusion furnace reaction. The symphonic sounds of the vibrating curved strings move from a speedy presto to joyous allegretto to the slower andante, filling this creative scene with awesome wonder and peace. In rapt attentiveness, from a vantage point within this lumpy island and close to this birthing cloud, I gaze at this magnificent explosion of light and radiation flowing swiftly from this newly luminous sphere. It is early on January 5, about 200 million years into creation history (myich).

Quickly, I move outside of this massive island of clouds to a distant viewing site, using my visionary gift of travel. At first, I see only this one tiny light from this first nuclear ignition. But as the day progresses, there are ten, then hundreds, thousands, and millions of lights flickering in the dark womb of the universe. Within the multitudinous clouds in this island in front of me, new nuclear ignitions and the emission of light and radiation are occurring.

In my vision, I am able to traverse the slowly expanding universe to take an inventory of these new luminaries. I move in the darkness and emptiness of space-time through its vast oceanlike expanses. I move close to clusters of clouds in another cosmic island of lumpy gas peopled by the hydrogen and helium families. I begin to explore this huge island of millions of pregnant clouds. Tens of thousands of nuclear furnaces within the families of clouds have ignited and are shining brightly. One massive cloud is the nursery for thousands upon thousands of new bodies of light. I notice that some bodies of light are gigantic in size, others moderate, still others quite small. Most of these young luminaries are spherical in shape and blue-white in color. The very large gaseous spheres tend to be more yellow in color.

I continue to move majestically through the huge, lonely ocean like stretches of black space-time. I visit the millions upon millions of lumpy gaseous islands. They continue to lure unattached orange hydrogen and golden helium 4 nuclei into their grasps to feed their numberless nurseries of pregnant clouds. One by one, parts of these clouds begin to coalesce, spin out, and, over time, bring forth nuclear fusion ignition, and millions upon millions of new lights are visible. As viewed at night from a mythical high-flying airplane, the universe looks like a vast sea of island nations consisting of hundreds of millions of large, lighted cities.

* * *

January 10, 400 myich, dawns. I continue to traverse on a curved path the almost endlessness of black space-time at breakneck speed. The universe has grown over these past five days. The clusters of lights have grown in number. The individual lights now number into the billions. The numberless neutrinos and photons are carrying the light and invisible but colorful radiation of these suns to my eyes with lightning speed.

Looming before me are other vast clusters of lights. I observe a primitive spiral cluster of lights, accompanied by multitudinous, kaleidoscopic-colored gaseous clouds. I say primitive because it has been formed by the collisions of a few of the gaseous lumpy islands that dwelled in the same local area. The imperfect spiral's huge disk contains a core of millions of lights rotating like a gigantic Ferris wheel, dragging three massive tails or arms filled with hundreds of millions of newborn luminous bodies. In the center of its central disk, I notice an area of intense activity. There appears to be some very intense but mysterious force that is whirlpooling near the entrance of a black opening. I report only what I see from a distance.

As I continue to traverse this spiral cluster's three arms, I see and hear the millions of lights rotating gracefully to the melody of a Strauss waltz orchestrated by the melodious strings and loops of strings, subatomic particles, making up the two atom families that inhabit this lighted body.

I see and hear Ruah in conversation with Elohim and Sophia within this spiraling wonder. Dabar is nearby.

"Quarkie was prophetic in saying that the golden helium 4 nuclei and the orange hydrogen atoms were about to give birth to new creations!" Ruah is commenting on the magnificent light show on display in the womb of the universe.

"She and her colleagues have energetically worked to ignite the first golden yellow luminary in creation history." Sophia pauses. Then she continues. "Their example and spirit have prompted the other families of orange hydrogen atoms and gold helium 4 nuclei in the billions of other gaseous islands to labor to give birth to similar nuclear fusion and ignition reactions. Look what their efforts have produced—this stupendous light show."

Elohim comments, "We need to give a name to these numberless island clusters of light."

"At a distance, these lights seem milky white." Dabar describes how the clusters of light in this nascent universe appear. "Let's use the name galaxy as a name to designate these new clusters of blue-white and yellow lights, similar to the spiral cluster over there. And let's call the individual light within the clusters by the name sun. Because of their brilliant sparkle, we can also give each sun another name: star."

"Excellent names!" Sophia is the first to affirm their appropriateness. "What about the enormous, multitudinous clouds that serve as nurseries for the billions upon billions of suns?" She pauses. After a time of musing, she says, "I name the individual cloud a nebula and many clouds nebulae."

"Nebula! Nebulae! They are appropriate names for these clouds of hydrogen and helium gases. And galaxy describes the island clusters of these gases to a *T*! And I like the name sun, or even star, to designate each point of light," Elohim adds. He looks in all directions in the bubble-womb of the black universe. He pauses to absorb its beauty. "Yet, I wonder what more exciting events are to come from these suns. I don't think that Quarkie was just talking about the creation of luminous suns, sparkling stars, huge nebulae, and gorgeous galaxies when she delivered her prophecy."

"We are just at the beginning of the universe's creativity." Ruah, who has inaugurated and birthed forth this universe, is also being prophetic. "Yes, we are still at the very beginning of this story."

"How exciting!" Sophia is full of anticipation.

"I wonder what Quarkie, Quarko, Quarkoff, and Hydro are planning next," Elohim says. "We must find them soon and talk with them."

Chapter 9
The Creativity of the Red
Star and Birth of Carbon

The black universe is alive with billions upon billions of lucent suns and billions of nascent galaxies on January 12, 500 myich.

I decide to explore in more detail the ebony womb of the universe and the creative work of their creators, the atom, nuclei, neutrino, and photon families. I again utilize my visionary gift of rapid arced travel within the universe bubble. I am able to exceed the speed of the photons.

As this magnificent womb of space-time expands like the uterus of a pregnant woman, I observe that two nearby lumpy, gaseous galaxies have collided. In this collision, they have created a new and enlarged galaxy of irregular shape. There are more and more suns igniting in the nebulae of all of the multitude of galaxies that inhabit this dark universe.

I observe clusters of galaxies moving away from me possessing a reddish tinge, while those moving toward me have a bluish tone. This observation helps me to distinguish the direction in which galaxies and their suns are moving in the universe. (From my earthly studies, I remember that this is caused by the Doppler effect. The receding electromagnetic light has a longer wavelength (red), and the same light coming toward me has a shorter wavelength (blue).)

From where I am standing, I see many clusters of galaxies. My eyes focus on a very distant spiral galaxy. In its vast rotating central disk, I see millions of suns. They are connected by the tug of gravity and the enveloping pressure of a halo of almost invisible dark matter and dark energy hovering over this galaxy. Each sun within the disk has its own orbit while moving synchronically with all the other suns in the disk.

As I gaze in wonderment, my eyes are captured by a spectacular, violent collision of two sun-clustered galaxies in the vicinity of the spiral galaxy. The crash melds two systems into a chaotic pattern. The impact releases vast quantities of intrastellar dust consisting of orange hydrogen atoms and golden helium 4 nuclei in addition to photons and radiation. This debris is swept up by nearby nebulae living in the galaxies. The force of the collision causes these newly seeded clouds to begin anew the process of sun formation.

I direct my gaze to another spiral galaxy. Its four long, rotating arms hold millions upon millions of stars in its embrace. Its mammoth disk and enormous arms resemble a monstrous, colorful pinwheel with tails. Again, I notice the mysterious force and dark opening in the center of its disk. As I peruse each massive arm, I am treated by the photon family to a light show. Along each arm, I see thousands of suns beginning to ignite. The nuclear fusion furnaces are igniting, illuminating the newborn round suns.

I observe numberless other island galaxies. Some are arrayed in elliptical forms, others in irregular shapes. As I look to the right, I am treated to the sight of a huge supercluster of galaxies containing billions of stars.

At noon on January 15, 600 myich, I look back toward the enormous disk of the spiral galaxy I had just passed. I see a very massive sun. Over a relatively short period of space-time, its yellow luster begins to take on a bright reddish color as its size rapidly expands. This red luminary is gigantic, about 100 million miles in diameter.

My special vision allows me to peer into the core of this reddish star to observe what is happening. Its core heats to more than 100 million degrees. This enormous temperature, along with the gravitational forces present, is pressing together the helium 4 nuclei. The birthing of a new element is taking place before my eyes. Trios of helium 4 nuclei are in the process of fusing to create this new element, which has six protons nuclei and six neutrons. It lacks accompanying electrons due to the extreme temperatures and pressures in the red giant's core.   Within just a few cosmic minutes, as I move from the core to its exterior, I see that this giant, red sun has halted its expansion. It appears to be a gigantic, red cinder. Suddenly, this expanded crimson mass begins to collapse onto its superheated core. Again, the interior temperature rises precipitously. The pressure of gravity, the action of the weak force, and the weight of the collapsing outer atmosphere are exerting mammoth pressure on the hydrogen, now hydrogen nuclei, and helium nuclei. Some of the newly created element, having six proton nuclei and six neutrons, are melding with one additional helium 4 nuclei to birth yet another mysterious new element made up of eight proton nuclei and eight neutrons.

At this crucial moment in the history of the universe, Ruah is present, along with Elohim, Sophia, and Dabar. Sophia is searching for Quarkie and her mates and Hydro.

She finds Quarkie in the white-hot core of this collapsing red giant. She appears as part of a new element.

Quarkie begins the conversation. She is exultant, filled with gratitude. "Greetings, Sophia, Elohim, and Dabar, my dear friends." She pauses and then asks, "Where is Ruah?"

"I am above you, Quarkie!" Ruah, the gentle, ghostly, godmotherlike figure, reassures her of her presence.

"Mysterious! I mean the experience through which my friends and I have just lived. The temperature at the core of our star began to increase higher and higher." She is excited as she recounts her story. "My helium 4 nuclei soul mates, Quarko and Quarkoff, and I met two other helium 4 nuclei. In this tremendously hot environment, we all felt a deep attraction to these helium 4 family members. This attraction was powerful. All the four forces of love energy, gravity, weak and strong forces and electromagnetism were present and operating."

Quarko affirms Quarkie's tale. "Yes, it was a profound feeling we experienced. It was like we were drawn to a special destiny. Yet, we were freely entering into this moment of dark new creation—this new union with the two other helium 4 nuclei."

"We are deeply grateful to all of you for empowering us to accomplish this new creative mission." Quarkie's gratitude exudes her person.

"I was nearby, accompanying you." Ruah wants to affirm her support of their creative mission. "You were free to enter this union or not enter it, just as you said, Quarko."

"Yes, it was our free decision." Quarkoff affirms his empowerment to decide. "But your encouragement was crucial in following our gut feelings."

"We applaud you, Quarkie, Quarko, and Quarkoff." Sophia commends the free use of their empowerment. "Your decision ended in the creation of a totally new element."

Elohim is quick to ask, "What name will we give to this new being?" There is a momentary pause as the question is pondered by the four Radiant Beings. Dabar finally breaks the silence. "Quarkie, you described your experience as dark and mysterious." A widening grin overtakes Dabar's godly countenance. He continues. "We will call this new black element a carbon nuclei, because it appeared out of the darkness and mystery surrounding your decision."

Sophia is quick to add to Dabar's definition. "Your new carbon nucleus is made up of six positively charged protons and six chargeless neutrons. Some time in the future, as you realize from your past experience, it will be joined to six negatively charged amber electrons to constitute a black carbon atom." Sophia pauses to savor this uplifting, vibrant scene.

Meanwhile, Elohim is looking around the white-hot fusion furnace. He asks, "Where is Hydro?" Elohim wonders aloud if Hydro has gone through some transformation into a new element.

Chapter 10
Hydro's New Friend and the Creation of Oxygen

"I found our friend, Hydro!" Elohim looks relieved. "He is living in a family of hydrogen nuclei that has given birth to the second element that was created after the black carbon nucleus."

It is about 1:00 pm, January 15, 600 myich. While this conversation is in progress, the gigantic outer shell of hydrogen gases is imploding and collapsing on this red sun's core. Quarkie, Quarkoff, and Quarko are nearby and listening to the conversation of Hydro and the Divine Beings. The latter are aware of their presence.

In a motherly way, Sophia approaches Hydro. "Hydro, my dear. We last saw you in that gaseous sphere which became the first sun to ignite and shine in our universe. What has happened to you in the interim?" She hesitates to allow Hydro to relate his story.

"What is a sun, Sophia?" Hydro asks. "I have never heard of a sun."

Sophia apologizes for the oversight. "We are sorry for this oversight. Out of your hearing, we, the Divine Beings, gave names to both the round spheres of light of which you are a part and the gaseous lumpy islands on which you live in space-time."

Elohim inserts himself into the conversation. "We named the lumpy islands galaxies, Hydro. And the round sphere of light where you live now we named a sun. It can also be called a star. We named the cloud in which your sun was born a nebula." After a brief pause, he adds, "I hope this answers your question."

"Please go on with your story, dear Hydro," Sophia urges.

Hydro begins his tale. "Well, our family of hydrogen atoms lost our electron mates as we descended deeper toward the white-hot core of helium 4 nuclei. We became naked protons." There is excitement and pride in his voice. "Gravity and an energy source drew us deeper and deeper into the inner conflagration at the core. I was directing the activity. While moving toward the core, one of my family friends, a female hydrogen nucleus, changed her identity by uniting her one proton with a nearby free neutron to form a nucleus of heavy hydrogen. As I watched her, the extreme temperature and pressure of the fusion furnace brought her and her companion into close association with another nucleus of heavy hydrogen. They bonded to become a helium 4 nucleus, possessing two protons and two neutrons." He hesitates briefly. "I decided to remain a light hydrogen nucleus living here on the fringe of the very hot core rather than follow my friend's course of action."

"My dear Hydro!" Sophia addresses him. "Would you introduce us to your friend? We haven't had the privilege of meeting her." Sophia is delighted to have the opportunity to meet yet another of the universe's creations. "We would also like to give her a name."

Since she has changed her identity, Hydro tries to locate her amidst this blazing nuclear furnace. "Lovely friend," he shouts. "I want to introduce you to our Creator-Beings!" When he finally spots her near the edge of the red star's core, Hydro notices that she appears perplexed. She is surprised and shocked by her new identity. Hydro empathizes with her. "What's wrong, my dear one?"

"I am lost. I feel disoriented." She appears very stunned. "What has happened to me?"

Sophia senses her confusion. "Dear friend, I am one of the Divine Beings who birthed this universe in which you exist. Let me introduce you to my husband, Elohim, our son, Dabar, and Ruah, the personal love who unites us as one. We gave birth to the universe where you live. In the beginning, we empowered the first particles and photons, who later became families of hydrogen and helium, to become creators. You have already become a creator of a new being."

Elohim then describes what happened to her and how she gained a new identity. "This is what happened to you. With your hydrogen family, you used your empowerment to create a new element. First, in a three-stage process, you bonded to a free neutron to form a heavy hydrogen nucleus. Then, you entered into a loving relationship with another heavy hydrogen nucleus to create helium 4." Elohim commends her for her courage and teamwork. "But the end. With your mates, you sought out and bonded with a black carbon nucleus. You have helped to create a new element consisting of eight proton nuclei and eight neutrons. You assumed a new identity. Don't be dejected. You continue to live but as an important partner in this new creation. You are a creator. We have endowed you with this power to create new elements. Welcome to your new family!"

"I am beginning to understand, friend Elohim." She reflects on her recent experience and begins to sound more joyous and hopeful. "I look forward to being a creator, oh Divine Beings." Her confusion is being banished. "Maybe it was a fatal attraction! I mean the intense energy of love that attracted me and my mates into this deep union with this dark, mysterious being, the carbon nucleus. We mated freely to become a new element." Her dejection is gradually dissipating as she recounts her story. "I love our

47

intimate relationship together. I now know who I am. Thank you, oh dear Elohim, for helping me understand."

"You now understand your recent odyssey, my dear one." Ruah is affirming her. "With all the changes you went through, no wonder you felt fearful and puzzled."

This female being is grateful for Ruah's empathy. Then she asks, "What name will you give to my new identity, oh Divine Beings?"

Elohim intrudes and says, "Before we give a name to your new identity, we need to give you a name." He hesitates. Everyone centers their attention on him. "I name you Hydrojean."

"I like that name, oh loving Elohim." Hydrojean is truly grateful for her name. She savors her name.

"Now, as for your new identity. I have an intuition about that, Hydrojean." Sophia's glistening gaze is fixed on her. "I intuit that in the future, you, Hydrojean, in your new existence will be extremely sharp and full of explosive power." Sophia draws a breath. "I name your new element oxygen. You are composed of eight positively charged protons nuclei and eight uncharged neutrons. Sometime in the future, you will be joined by eight negatively charged amber electrons and will be called an oxygen atom. You will live in this new family, oxygen."

Elohim decides to do some introductions. "Hydrojean, I want you to meet some new friends, Quarkie, and her soul mates, Quarko and Quarkoff. They live now in this nearby carbon atom. They also have gone through an evolution. In the first second of creation, they were birthed, the first particles in the new creation. Within a very few seconds, Quarkie, Quarko, and Quarkoff met in space-time. Drawn by the energy of love, they united to form one of the first stable particles, the proton. They were shortly united by a strong energy force to a free neutron particle. Within a few moments, through a three-stage process, they joined with another proton-neutron element to become helium 4. So this threesome are members of the helium 4 nuclei family, now joined with two other family members to create the first carbon atom. For a short time, you belonged to the helium family as they did."

Quarkie has been listening with intensity to Elohim's description of their creation and evolution. She begins to address Hydrojean. "Hydrojean, my name is Quarkie."

"So happy to meet you, Quarkie." Hydrojean is genuinely glad to meet her.

Hydro feels neglected. With anger and jealousy oozing out of him, he blurts out, "Have you forgotten about me?" He is peeved at the Divine Beings. He spews out his indignation. "You came searching for me. You find me. I introduce you to my female friend. You proceed to ignore me."

Sophia understands his anger and feeling of rejection. But she also realizes that these feelings are arising from his pride and egoism. She wants to calm him and make him feel accepted. "My dear friend, Hydro. You have a right to feel this way. We did ignore you. Please forgive us." Sophia is truly contrite. Hydro appears to accept her apology. She continues. "We appreciate your introducing us to Hydrojean." Hydro begins to feel more appreciated and accepted by all. "Hydro, let me introduce you also to Quarkie, Quarko, and Quarkoff. They live in this new being, carbon."

"So good to meet you, Hydro." Quarkoff speaks for the trio.

Hydro again feels a part of this conversation. Feeling accepted, he says, "Continue your story, Quarkie."

Quarkie picks up the story where Elohim left off. "Hydrojean and Hydro, fate has brought us together in this same gaseous nebula that has given birth to this new sun. The rotation and gravitational pressure of our cloud drove us into its core. Recently, as we began to run out of fuel, we learned from members of your hydrogen family, being pushed toward this core, that our sun has expanded and changed its color from yellow to red. They informed us that the exterior of our red sun has begun to collapse. The temperature has risen to hundreds of millions of degrees. Our helium 4 nuclei felt impelled by an intense energy of love to bond with two other helium 4 nuclei. Thus, we birthed a new element, which Dabar has called carbon. I feel so privileged to live in this new creation. You, Hydro, were important to the creation of our new identity because you were supervising the effort to fuel our nuclear furnace. You urged your family members to unite to produce more helium 4 nuclei fuel through a three-step process."

Hydro is appreciative that Quarkie recognizes his efforts. He feels some kinship with her and her friends. "Do you have the same feeling that I have, my quark friends, that this is just the very beginning of a long, momentous journey? I look forward to continuing to be a leader in this odyssey of creation." Now Hydro again looks assured and self-confident.

"Indeed, we do, friend Hydro. There is much, much more to our expanding journey in the womb of our universe. We are truly privileged to be a part of this odyssey," Quarkie concludes.

I am full of wonderment at what I have just visioned. Yes, I feel sure there is more to come of a surprising nature. What will happen to this robust, red sun that is collapsing on its superheated core? And Hydro, where will his pride and egoism lead him? To greater participation in the creative journey of the universe or to some dark fate!

Chapter 11
The Collapse of the Red Star

"What an immense red sun!" I shout in the emptiness of the space-womb of the universe. I have just left the core of this red giant after being present to the conversations between the Divine Beings and some of the members of the new element families, Quarkie, Quarko, Quarkoff, Hydro, and Hydrojean. It is early evening on January 15, still about 600 myich.

I am flabbergasted that the once gigantic bright red sun is continuing to collapse upon its tremendously hot helium 4 nuclei core.

I recall a remark that I overheard between Ruah and Dabar just a few days ago. Ruah said to Dabar, "Notice that the more massive the sun, the faster it burns its hydrogen and helium fuel and possibly the quicker its life ends. And oh, what a dramatic end it may be, Dabar!"

At the time, Ruah's prediction intrigued me greatly. I thought to myself, *the quicker and more massive a sun grows, the more dramatic may be its demise!* I am now musing this sun's fate. *How will it end?*

I watch the red sun's collapse but want to experience what is transpiring at its core, where my friends are residing. With my visionary gift, I am again able to descend toward the core of this diminishing red giant without being incinerated and annihilated. As I again approach the white-hot core of the furnace, I hear an ongoing conversation. I eavesdrop.

Sophia is talking to Quarkie. "What are you experiencing now, Quarkie?" She senses the core's atmosphere is heating up.

Quarkie is filled with excitement. "Wow! Our red sun community is super-active and expectant again. There is tremendous pressure coming from above us. The hydrogen family is pressing in on those of us at the core, the helium 4, carbon, and oxygen families. The hydrogen atom family continues to decrease in numbers as they lose their electrons and bond to create helium 4 nuclei to fuel our core."

"What else is happening in the core of this furnace?" Dabar is inquisitive.

"The helium 4 family continues to fuse together to create more members of our carbon family! The latter are melding with an additional member of the helium 4 family to add to the oxygen family." Quarkie is vibrant and filled with enthusiasm. But her lively mood suddenly gives way to a concern and a question as she further describes the situation at the core of the sun's furnace. "Like the hydrogen family, the helium 4 nuclei family is diminishing in numbers. Sooner or later, we will run out of helium 4 nuclei fuel in the core of this nuclear-fusion furnace." She pauses and muses, "With this fuel exhausted, I wonder: what will happen to this red sun?" First enthusiastic about the creative activity going on in the core of her sun home, she is now becoming pensive.

As usual, the fairy godmother, Ruah, is brooding over this intense scene. She gives an explanation. "The helium 4 nuclei, like you, Quarkie, Quarkoff, and Quarko, who have fueled the sun's furnace, have sacrificed in love to give birth to new element families, nuclei of carbon and oxygen. Do you notice that the temperature is increasing toward many billions of degrees, my friends? This red sun is swiftly collapsing and soon will implode onto its core, where you are now, causing the temperature to soar." Ruah pauses momentarily.

Quarkie, Quarkoff, and Quarko become more anxious awaiting the conclusion of Ruah's explanation. Ruah looks directly at them. "As just happened with carbon and oxygen nuclei, this may make it possible for those of you in the hydrogen, helium, carbon, and oxygen families to open yourselves to the increased pressure of gravity in order to meld together in love and give birth to additional families of elements."

"But, Ruah, by losing our identities as helium 4 and hydrogen, and later, carbon and oxygen, aren't we losing our connections?" Quarkoff is expressing his deep concern. "What about our being a web of creation?" He looks intently at Ruah.

"Remember your basic connection, gravity!" Ruah emphatically concludes her response to Quarkoff. "You may change your identity many, many times, but gravity joins you to the larger community within the web of creation. Gravity also communicates our love to all of you, believe and trust me, dear Quarkoff."

"Thank you for your reassurance, Ruah!" I sense Quarkoff's relief due to Ruah's reassuring statement.

Hydro is on the limits of the conversation. He enters the interchange with a question. "Ruah, what comes next?"

"Your fusion furnace is gradually exhausting its golden, hot fuel. Once the imploding outer surface and inner elements bombard this core, the temperature will reach massive intensity. In this hot, energized environment, as we have already mentioned, you and your element friends may give birth to even more exotic, heavier element families." Ruah pauses briefly and then continues. "Within just a few moments, Hydro, Quarkoff, Quarko, and Quarkie, I intuit that something stupendous is going to happen!"

"What do you mean, stupendous, Ruah?" They shout together in deep fear and trepidation. They look at each other, stunned.

They wonder if the impending catastrophic event will be for their benefit or their demise.

"Trust us and trust the future, my dear friends!" Sophia, who had initiated the conversation, says with loving conviction. Then she, Ruah, and Dabar vanish.

Hydro, Quarkoff, Quarko, and Quarkie stand in awe, anxiety, fear, and expectation.

Chapter 12
The Red Sun Explodes

The once bloated red star continues its collapse. The descending hydrogen atoms are crushing together all of the gaseous particles in its path. The immense weight of all of this mass comes crashing down upon the blazing core of fusing nuclei. Its pressure crushes the various elements together. The gigantic heat, pressure, and density cause the birthing and melding of new and heavier families of atoms. Since they are lacking accompanying electrons at this time, they are nuclei of these elements.

My vision gift helps me to move from the core of the collapsing red star to a vantage point a long distance from it.

Suddenly, the remnants of this red star at its core explode. Its crimson luminosity obscures temporarily my view of much of the dark universe. The bursting red star paints the nearby space-time in brilliant shades of reds, blues, greens, and yellows. The subatomic strings and loops of strings within this exploding mass vibrate and resonate to the strident tones and beat of the finale of Tchaikovsky's *1812 Overture*. The burst of the canons in this musical setting accompanies the red sun's explosion. In all directions, the newly created hydrogen, helium, carbon, oxygen and all of the heavier elements, microscopic in size, are blown violently into nearby intragalactic space. As they travel outward at fantastic speed, they join themselves to free electrons to become atoms. Some of these atoms are captured by the nearby nebular clouds within this galaxy. Others escape this spiral galaxy and begin long, curved journeys through the emptiness of black space-time. Some will end their journey in other galaxies.

The colors in this explosive event far surpass the most gorgeous sunset I have ever witnessed or imagined. Wafts, whirls, and waves of lemon, saffron, pink, aqua, turquoise, violet, and ruby clouds commingle and rush outward. My eyes follow the flight of this multitudinous primeval atomic dust rocketed into intragalactic and intergalactic space-time.

To gain a further perspective, I travel a vast distance into space-time in the vicinity of another galaxy and look back on this scene. From this vantage point, this huge cataclysm appears to be the initial burst and the glow of a small firecracker, which gradually dies down.

I am amused by an extraordinary scene being played out before me. Elohim and Sophia are having an animated conversation with Quarkie, his proton mates, Quarko and Quarkoff, Hydro and two other new characters as they are shot out into intragalactic space-time. Again, with my visionary gifts, I have the ability to catch up quickly to these personages and accompany them.

As I reach them, Quarkie is expressing her feelings and thoughts about the implosion and explosion that has devastated their former home, the gigantic red sun. "The implosive pressure crushed us into a closeness I have never experienced before." She is excited as she speaks. "I felt a profound affinity to all the atom family members, even the new ones. I have never experienced such depth of feeling previously. There was a frenzy of activity and a loving interchange that energized all of us. Yes, it was an awesome experience."

"Just before this explosive event, you, Quarkie, and your friends were experiencing ambivalent feelings of awe and anxiety." Sophia is probing her experiences with sensitivity. "You seemed to have moved through these complex feelings to courageous activity."

Quarkie listens carefully to Sophia and responds. "Do you remember my initial fears after my birth in the first second of creation?" Quarkie continues her reminiscence. "When I decided to join Quarko and Quarkoff to create the first proton, I did so with foreboding and trepidation. But when I made the leap of friendship with my proton mates and joined them in love and creativity, I learned a great lesson." Quarkie pauses. She is filled with deep emotions. It is apparent in her quivering voice.

Sophia empathizes with her. Then she asks, "My beloved Quarkie, what was the lesson?"

"I learned to trust you, Sophia, Elohim, Ruah, and Dabar, to trust myself and my companions, and to trust the future!" Quarkie begins to relax as she proceeds. "In fact, those were Ruah's last words to us before the implosion and explosion of our red sun home: 'Trust us and trust the future!' And I do deeply trust you and the future, my Divine Friends."

Hydro is cynical about the future. "I played only a minor role in the recent creativity and destruction of my red sun homeland. I was part of the hydrogen family being drawn by gravity to press down on the helium 4 family. Living near the outer limits of the reddish sun, I did not undergo an identity change and crisis like Hydrojean did. I am glad I did not. Viewing this dynamic scene from a distance, I have become skeptical about the future of the universe."

"What are you trying to say about the future?" Elohim is inquisitive.

"I don't buy into Quarkie's naive feeling of trust about the future." There is a tone of arrogance in his voice. He continues. "Your constant sugary emphasis on the love that exists among our atom families repels me, Quarkie. And your simple-minded trust in the future nauseates me!"

Quarkie is shocked by Hydro's pointed criticism of her. Anger at Hydro wells up within her. His stinging criticism fortifies rather than diminishes her feelings of trust. "Okay, you do not trust in the future like I do. You call me naive." She halts momentarily and ponders a further response to Hydro's cynicism. She continues. "Let me ask you pointedly, Hydro." She goes right to the heart of this discussion. "Do you, Hydro, trust in our Divine Friends, Sophia, Elohim, Ruah, and Dabar?"

Hydro feels profoundly challenged by Quarkie. He muses over this pointed question for some time before responding. "I ... I ... I do not fully trust them."

"Why; Hydro? Why?" Quarkie pursues Hydro with persistence. "Why do not you trust our Divine Friends?"

"They are too vague about the future. If they are so almighty and all-knowing so as to be our originators, they should be able to give us at least a glimpse and synopsis of what events lay in the future." As he speaks, Hydro comes in touch with some deeper disdain of Elohim, Sophia, Ruah, and Dabar. "I put more trust in myself and less trust in our all-powerful Divine Beings."

"We love you, Hydro! We care for you deeply! You are intimately involved in this creative process. We have empowered you to be a creator." Sophia exudes sincere compassion and deep empathy for Hydro.

"I don't like the uncertainties of this creative process." Hydro is trying to be truthful. "How can I trust you when you don't know the outcome of this whole creative process?" He muses for a few moments in silence. "I just cannot trust you, so-called Almighty Ones. I must part company with you and go my own way."

The Divine Beings are stunned by his abruptness and rudeness. Quarkie, Quarkoff, and Quarko can hardly believe what they are hearing. They are angry at him.

Elohim is the first to respond to Hydro's harsh remarks and angry decision. "Yes, Hydro, we still love you. If you would want to speak to us at some future date, rest assured that we are present in the universe to dialogue with you. Just call for us and we will be with you. We wish you well as you depart. You will continue to be a part of the creative process, Hydro, but as always, you are free to make your own decisions."

Hydro turns and leaves them. He moves out into the mysterious black regions of space-time.

Sadness envelops Elohim and Sophia and the quark trio. They are silent for a long time as they too journey together through the vastness of dark space-time.

Quarkoff breaks the silence. "Divine Friends, what name do you give to this momentous implosion and explosion we have just experienced?"

Sophia is the first to respond. "Quarkoff, this is a new event—a nova. In addition, it was a super happening. It is rightfully called a supernova."

"Yes, the explosion of our massive red sun was a supernova." Quarkoff is filled with elation. He begins to overcome the sadness of Hydro's confrontation and departure. "And we, Quarkie, Quarko, and I, were an intimate part of this creative event that gave existence to so many new families of atoms. Divine Beings, you must meet some of our new friends. We met them in the midst of all the calamitous activity before and after the red sun's violent explosion."

"We would be very happy to meet them, Quarkoff." Elohim is inviting some introductions.

"Let me introduce you to a member of this new family." Quarkoff is proud to introduce this first member of a new family. He proceeds with his introductions. "My next new friend is part of another new family."

Quarko interjects himself into the conversation. "Elohim, Sophia, you need to give names to our friends, as you did for us." He then asks, "What names will you give them?"

Elohim ponders briefly. "Your first new friend and all of his family members, Quarkoff, will be called Nitrogen Atoms. Let us shorten your personal name, new friend, and call you Nitro. As with members of the oxygen family, the *gen* of your name means the ability to give life. In the future, you are destined to join your family members and all of the other atom families in some very momentous creative processes, I feel sure."

Elohim nods to Sophia, who understands what he means. "I am to name the family of the second new friend. You, our new friend, and your family members, will be an element that will be light-bearing. You have a future full of light and life in the expanded universe. You will be called Phosphorus atoms. You, phosphorus friend, will bear the name Phosie. Welcome! Take your place among the web of creation."

"Thank you for my new name, Nitro!" He pronounces it proudly. "Quarkie told us about the beginnings of creation and how she first met you. She was overwhelmed by your sensitivity to her initial fears." Full of sincerity, Nitro proceeds. "I will be proud to be a part of the web of creation and the creative process of the universe."

Phosie is the next to speak. "Sophia and Elohim, I am grateful for your loving welcome to Nitro and me." She pauses briefly and smiles. "What a privilege to be a part of the creation of this universe. I am ready to bear my light in order to help to create the universe."

"You will both have your opportunities to be creators. But you must trust us and yourselves and trust the future." Elohim finishes his short remarks.

I sense a deep peace envelop this part of the fast-moving, multicolored community of atoms. The whole scene seems unbelievable. All the characters in this drama are shooting out into space-time at fantastic speed. I believe that this explosive event of the first supernova will be very important for the future of the creative processes of the universe.

What will be the next creative event in this fantastic journey?

Chapter 13
Hydro Challenges His Atom Friends

From early February, 1.2 byich, to May 1, 5.0 byich, I am able to watch the universe continue to expand. Against the black background of the universe, as I look in all directions, the sky continues its radiant light show. Hundreds of millions of suns multiply and grow into the hundreds of billions. Within each galaxy some huge, fast-growing yellow suns develop over a relatively short period of time. When they use up all of their helium 4 nuclei fuel, they implode and explode, seeding their local communities and the black bubble of the universe with the ninety-two light, medium, and heavy atom families.

How can I describe the immensity of the expanding universe? (From my earthly life, I recall that our sun is 93 million miles from the Earth. The closest star to our Earth is estimated to be 4.5 light years away. Each light year is the equivalent of 6 trillion miles traveled by photons speeding about 186,262 miles per second. Thus, this nearby star is about 27 trillion miles from the earth. If these distances boggle our ability to understand the massive distances of space between our Earth and our sun and closest star, think about the super, super-gigantic 5 billion light years' diameter of the bubble-womb of the universe in early May, 5.0 byich. These distances are literally inconceivable to our human minds.)

All through this space-womb, there are enormous galaxies; some are spiral, others round, still others elliptical or irregular.

There are superclusters of smaller galaxies making up unbelievably huge ones. Yet some are moderate and others small in size. Yet even the smaller galaxies are made up of millions of older yellow suns, hundreds of millions of newly formed blue-white baby suns. All the galaxies with their multitudes of suns have more than sufficient room to rotate on their axes around the center of their galaxy with rarely the suns or galaxies colliding into each other. (Again, to attempt to understand the distances between stars within a galaxy, I return to my earthly experience. Our sun is approximately one hundred times larger than our Earth. I see suns that are five hundred times the size of our sun. Such huge suns are fifty thousand times the size of our Earth. Within a galaxy, this is just one sun among the billions of suns living there.)

The blackness of space in between these hundreds of billions of galaxies is awesome. As I stand in the middle of this vast, desert-like expanse of space-time, the billions upon billions of suns look like tiny pinpoints of light. An occasional implosion and explosion of a supernova intensifies the light from a pinpoint of light to the flare of a firecracker. The only travelers in this black desert are dust particles of heavy and light atoms, yellow photons, emerald neutrinos, and an occasional rose, yellow, or blue stray gaseous nebula or yellow pilgrim sun.

Close to a supercluster of galaxies, but still situated in the blackness of space-time, I come upon some friends. (I should confess that they are not personal friends. I know them only through this vision. They are unaware of my presence. Yet I still have a gut feeling that I am related to them in some intimate way.) Present are the trio, Quarkie, Quarkoff, and Quarko, now members of the carbon family; Hydrojean, now a member of the oxygen family; and Hydro. Nitro, and Phosie are also present. They are involved in a serious conversation.

"I still don't trust our Divine Beings." Hydro is still defiant. He directs his words to Quarkie. "Quarkie, you are a simpleton," he adds with sarcasm and in ridicule. "You said to your Divine Friends, 'I do trust you and I trust the future.' Can you really trust them to lead you to a secure future, my foolish friend?"

Quarkie absorbs Hydro's personal criticism, then dismisses it immediately. "Hydro, you are my friend and pilgrim with us on our creative journey. I believe deeply that you are making a grave mistake by insisting on being independent. Don't you realize that the entire atom families are one diverse community bound together in love by gravity? We are a vast web of creation. Elohim, Sophia, Ruah, and Dabar have empowered us to be creators. What a sublime mission."

Hydro is ready to meet her challenge. "You have naively swallowed their propaganda. If they are so powerful and all-knowing, why can't they detail how this journey will end? You are fools to follow leaders who have no detailed plan for creation."

Quarkoff takes up this challenge. "Friend Hydro, our Divine Beings trust us. They have empowered us to make decisions about how the course of creation will go. This instills confidence in me. I like their empowerment of me, Hydro."

Quarkie affirms Quarkoff's experience. She takes up the debate. "I too feel good about our empowerment. We are given decision-making power. Yet, at crucial moments of choosing between alternatives, I always sense Ruah hovering over us radiating love and wisdom. She does not make the choice but energizes us to be able to make good choices." She looks compassionately upon Hydro.

"Again, I press you, Quarkie and Quarkoff." Hydro becomes more logical and less sarcastic as he says, "Where is the overall plan for this creative activity of the Divine Beings? What is the end toward which they want us to strive?" Hydro looks intently at the quark trio.

"The overall plan we are following is trust and empowerment. I believe deeply in this plan. I trust our Divine Friends." Quarkoff is very emphatic.

Though silent up to this time, Quarko enters the dialogue. "Hydro, I deeply support this creative plan. I trust in this plan and I trust in our Divine Friends."

"You blind fools! Trust ... empowerment!" A venomous spirit reenters Hydro. "They do not constitute a plan! You are misguided! You are stupid!" Hydro's voice is ringing with anger. "I trust in myself! I empower myself! Enough of this ridiculous conversation. I must go to plan my own future." Defiant, he moves away from the trio.

I am startled by Hydro's strident attitude of self-centeredness and the strength of his defiance. Where will this spirit of independence and defiance lead him? Only our friend, space-time, will tell us!

Chapter 14
Proud Hydro Looks to Take Leadership of the Creative Process

I am situated close to the center of a huge, spiral galaxy rotating clockwise. It is May 5, 5.2 byich. I cannot miss seeing the Divine Quaternity. They are in loving conversation. The subject is Hydro.

Sophia expresses her deep, motherly concern. "Hydro continues to assert his independence from his hydrogen family and the other atom families. He does not understand our empowerment of the proliferating families of beings in the universe."

Dabar is equally concerned. "He wants us to outline a complete plan for the future of the universe. We are capable of making such a comprehensive blueprint and of carrying through with its accomplishment." He ponders briefly before continuing. "But that would make our creations powerless. They would be enslaved to our minute direction of their activities."

Elohim finishes Dabar's thought. "And from the beginning of our birthing of this universe, we have empowered Quarkie, Quarko, Quarkoff, Photie, Hydro, Hydrojean, Nitro, Phosie, and all of our ninety-two families of atoms to be partners in our creative work. That has been our plan."

"I have been hovering over them with our creative energy of love to energize, enlighten, and encourage them." Ruah outlines her part in the creative process. "I have talked to Hydro, just recently. But he is still determined to be independent. He feels we are incompetent. He believes our plan will lead the universe to chaos and destruction."

"How can we help Hydro to trust us and our present plan for the future?" Sophia loves Hydro so deeply, even in his defiance of them. "We must warn him that the universe's law is one of community and cooperative effort. His deviation from this law could lead to his destruction."

Elohim is following the dialogue. "Hydro has so many talents. If only the energy of his independent spirit could be harnessed and directed toward creative work within the community and the web of creation." Elohim has been devising a plan as the dialogue had gone on. "Let us send Dabar to counsel with him. Maybe with Dabar's help, Hydro can change his mind and independent spirit to again join our creative community."

Dabar is quick to respond. "Yes, I will undertake this mission. Stay close to Hydro and me, Ruah, when we meet. Hydro desperately needs your energy of love and wisdom."

"I will be close to both of you, Dabar, when you converse. Yet, we know that he has the freedom to continue to defy our loving concern." Ruah pledges her support but also reminds the other Divine Beings of the personal law of freedom, which is a part of the whole creative process.

As the interchange concludes, I understand more profoundly the Divine Beings' cooperative, creative plan for the universe. I understand that they love and care for Hydro, even in his open defiance of them. I wonder how Hydro will respond to Dabar.

* * *

A carousel of suns looms before me! Amazing! It is a round-shaped galaxy containing hundreds of millions of suns. I am using my visionary gift to traverse this galaxy swiftly. Let me describe its origin and workings.

This circle galaxy in which I am traveling began to form from a lumpy island about January 5, 200 myich. It consisted of hydrogen atoms and nuclei, helium 4 nuclei, free protons, neutrons and electrons, photons and neutrinos. They were arrayed in millions of multicolored gaseous clouds. At its beginnings, the multitudes of nebulae in this round galaxy consisted of about 80 percent hydrogen and 20 percent helium.

Now, on May 5, 5.2 byich, and after numerous supernovae explosions in various areas of the galaxy, 1 percent of the elements in these galactic clouds consists of light and heavy members of the ninety-two atom families born out of these explosions.

These intragalactic gaseous nebulae are the nurseries for new sun development. Nebulae were and still are objects of exquisite beauty and gigantic size. The wafts of pink, red, yellow, green, and blue clouds eddy, whirl, wane, and mingle, creating a moving canvas of interacting, ever-changing, magnificent design. Wondrous gravity still attracts the hydrogen and helium 4 families. Like a skillful ice skater, one section within a nearby nebula begins to whirl on its axis to create a sphere, drawing the heavier helium 4 nuclei deeper into the core of this soon-to-be sun while the lighter hydrogen gases end up surrounding the core. The pull of gravity exerts pressure on the core. The hydrogen atoms go through a three-stage process to create more helium 4 nuclei to fuel the core while releasing neutrinos. The temperature in the core rises to millions of degrees. Suddenly, when the critical temperature reaches 18 million degrees, the helium 4 core ignites. A mushroom cloud of brilliant yellow light and spectral radiation, carried by the photons, rises and suffuses throughout the huge, spherical body. It bursts out beyond this sun into its nebulae nursery and beyond.

Thus, this particular nebula spins out a newborn sun. Suddenly, the universe also observes the appearance of this tiny new light in space-time.

The billions of suns in its gargantuan disk are rotating and orbiting around the core of this round galaxy. I notice that in the center of this galaxy, some nebulae, small stars, and elemental dust and gases are being captured and sucked into a dark cavity by a powerful but mysterious force that I had observed before. I move closer to look at this phenomenon. As these bodies pass a certain point, they are caught by a tornadolike current that draws them toward a dusky, murky aperture.

As I move to a point close to the center of the core of this immense galaxy, I feel an intense gravitational pull as a cascade of powerful X-rays is emitted from this dark, mysterious cavity.

* * *

It is May 10, 5.4 byich! While still traveling in this round galaxy, I see Hydro in the company of Hydrojean, Nitro, and Phosie. They are involved in an intense discussion. I recognize Hydro's voice.

"Nitro and Phosie, you are the newest members of the atom families." Hydro is emphatic. He tries to win these neophytes to his side. "Do not be duped by Hydrojean, Quarkie, Quarkoff, and Quarko. They and their families of atoms are blindly following the Deities." His tone becomes harsher, tinged with intense anger. "The latter have no credibility. They are planless about the future of this universe. They would have you believe that if you follow them, some great purpose will be ultimately achieved. This is nonsense. I urge you; do not be deceived. Follow me to independence, leadership, and great creativity in this universe!" Hydro finishes his sales pitch with promise of future reward and possible notoriety.

There is silence for about a minute. Nitro and Phosie are speechless. The threesome is in a temporary state of shock.

Haltingly, Hydrojean begins a response to Hydro's appeal to join him in his independence movement. "I disagree with your main premise, Hydro. I have believed from the beginning of my existence, when I initially met our dear Divine Beings, that their plan of furthering the future development of the universe was a good one. Why? Because the hydrogen family, of which we, Hydro, are a part, are empowered to be creators." Hydrojean is growing in confidence as she speaks. "I would rather be a vital part of the atom families, woven together as a web of creation, than to be a loner like you, Hydro. We all gain strength from our associations with the other members of our community, and especially with Elohim, Sophia, Dabar, and Ruah. They deeply love and respect us. I gain deep self-confidence from their loving trust in me." As she concludes, her self-assurance permeates her whole being.

Hydro turns to Nitro and Phosie. Being newer family members, he again believes they are vulnerable to his argument. He wants to enlist them to join him in his revolt. "Nitro and Phosie, I hope you have been spared the brainwashing that Hydrojean has undergone under pressure from the Divine Beings. Will you follow their foolhardy plan, or will you join me in independence from them and their ill-conceived scheme?"

I observe that Nitro and Phosie feel violated by Hydro's diatribe and questions. Yet, they feel insecure. Their experience with the Divine Beings has been very limited. Nitro is the first to respond. "Yes, I belong to one of the newer families of elements. But I agree with Hydrojean. Elohim, Sophia, Dabar, and Ruah have been very caring and sensitive to my needs." Nitro is also growing in self-assurance. "I like being part of a community and being empowered to be creators in our universe." He stands tall and confronts Hydro. "I will not join you, Hydro! I respect you, but I must speak the truth in love to you. You are too proud and self-centered. I believe you will end up destroying yourself." Nitro is reassured by his own remarks.

Phosie is buoyed up by Nitro's strong support of the Divine Beings and the other atom families. She also has trusted the Divine Beings from the beginning. She faces Hydro and says, "I agree entirely with Hydrojean and Nitro. I feel deeply respected by our atom communities and by the Divine Beings." She appeals to Hydro. "Give up your self-centered quest for power, Hydro. Rejoin our community and the web of creation. I want you to be a part of us! I love you, Hydro! Don't continue your individual course of action. I fear, too, that your willfulness will lead you to destruction." Compassion exudes from Phosie's being.

"Well, this conversation has been a complete waste of time!" Hydro is very vexed at this trio of atoms. He turns and abruptly departs. From a distance, he shouts, "You are stupid fools!"

I sense that Hydrojean, Nitro, and Phosie are profoundly saddened by Hydro's rejection of them. They truly fear for his future welfare. I, too, wonder what fate lies in wait for Hydro.

Chapter 15
Hydro's Fatal Temptation

"Hydro, could I speak to you?" Dabar had just caught up with him as they both approached just a short distance from the center core of this round galaxy. The date is early afternoon of May 10, 5.4 byich.

"I do not have time for your laughable, nonsensical arguments." Hydro speaks with firmness and arrogance. "You probably want to talk me out of my independent spirit in order to enslave me to your preposterous plan."

"I love you, Hydro," Dabar proclaims with candor and sincerity. "I feel sure that none of my arguments will dissuade you from your independent spirit and actions. But I do want to give you a warning."

"So this is your new stratagem." Hydro becomes wary of Dabar's motives. "You are going to frighten me in order to try to control me. I refuse to submit to your devious tactics." He is resolute in will.

"There is danger ahead, Hydro. Truly there is." Dabar pleads with him. "In the center of this massive galaxy is a dark, awesome aperture. Possibly you have seen them in other galaxies. It was created long ago by a massive sun coming to the end of its life and collapsing on itself. The pull of gravity on the surface of this stricken sun is so gigantic that almost nothing can escape its mighty grasp. If you pass a certain point from the center of this black monster, all surrounding bodies, stars, nebula, light, and elemental dust and gases are sucked up and thrown into its cavity to their total annihilation. We have named it a black hole. You must stay away from its lure and awesome power, Hydro."

"Balderdash, Dabar!" Hydro is defiant. He is calling Dabar's bluff, he reasons. "I can take care of myself. I do not need your protection, oh deluded Deity!" Sarcasm drips from his fiery words.

"Believe me, Hydro, in just a short time, you will face a supreme test between independence and annihilation." Dabar makes his last loving plea. He feels chagrined because of Hydro's intransigence.

Hydro sulks and moves away in angry silence.

I am overwhelmed by the exchange. Where will this drama take Hydro?

* * *

A little later on May 10, 5.4 byich, I transport myself to the vicinity of the black hole at the center of this round galaxy. It is a massive opening. The explosion of a supernova was its origin. (From my previous studies, I learned that a black hole is formed from the residual core of an imploding red sun, which has a mass exceeding three times the mass of a typical sun.)

I stay at a distance to avoid its awesome clutches. As Dabar said, the force of the gravity is overwhelming even at a safe distance from its awful grip. I see its dynamic gravity whirlpooling at the hole's entrance. I watch a small, older yellow sun being snatched at edge of its gravity field, being dragged relentlessly and then pulled apart at the lip of the black hole by its intoxicating power. As it enters the fearsome hole, its atoms are being heated, torn apart, and ground down into its elementary particles and swept into the depth of this hole of darkness. The power of gravity is so immense and the temperature so awesome that I see no light escaping from it. It does emit intense X-ray radiation with its spectrums giving off ultraviolet, blue, yellow, and red colors.

From a distance, I see both Hydro and Hydrojean moving in close proximity to the fearsome edge of this black hole and its tornadic whirlpool in intragalactic space-time. They are traveling at a very high speed.

"Can you feel that tremendous tug attracting us, Hydro?" Hydrojean is beginning to become frightened. She has never before experienced such a powerful attraction.

"That must be the black hole that Dabar warned me about recently." Hydro is curious and possibly beginning to act foolhardy. "I'm fascinated by its power, Hydrojean. I must investigate this mysterious black hole. If its tremendous power can be harnessed, I might be able to use it to further my own creative plan for the universe."

Hydrojean's anxiety grows. She becomes concerned about Hydro's brashness and how it might harm him. "If Dabar warned you about black holes, Hydro, please heed his warning. He cares for you. That's why he warned you."

But Hydro throws all caution to the wind. He veers toward the edges of its intense gravity field. Even Hydrojean, who is maintaining her course through the center of the galaxy, is experiencing its massive power. Her fear grows. She watches Hydro being drawn strongly toward its hellish hole. He has crossed the point of no return. The velocity of the black hole's whirlpool of gravity exceeds the speed of light. (I remember from my study of black holes that the edge at which the speed of gravity exceeds that of the speed of light is called the event horizon.) Hydro has crossed over this event horizon.

Hydro's mood begins to change from curiousness to daring. But suddenly it is dawning on him that he has entered a state of extreme foolishness. He begins to sense the drastic danger that he is approaching. The attraction of the black hole's funnel-like action clutches Hydro in its grasp. He is being drawn infallibly, against his free-spirited will, into the midst of its maelstrom. He is being faced with a reality at which he had scoffed. His spirit of independence and pride has brought him to the brink of destruction. He begins to rue his foolhardiness and careless dismissal of Dabar's warning. He looks to the future. He is beginning to realize he has permanently cut himself off from the future. He is in extreme anguish. He cries out for help. But there is no assistance that can save him from virtual extinction. His spirit of independence has cut him off from his atom families and their loving community and the Divine Beings. There is no salvation for him.

"How stupid I have been?" Hydro is in deep remorse. "I did not listen to Quarkie, Hydrojean, Phosie, or Nitro and their entreaty. I dismissed the loving counsel of Dabar, Elohim, Sophia, and Ruah. I am finished!"

He sees clearly the vortex of the black hole. He is losing consciousness as he is dragged through the entrance into its chamber of horrors.

My visionary gift allows me to enter this black hole but not be affected by its awesome power. I observe Hydro being torn apart into his elemental particles, proton, electron, then quarks, and finally strings and loops of strings. Before my eyes, in this vision, I see Hydro virtually annihilated.

I look around this black hole. All the mass drawn into its ruthless grasp is scrunched down into a point. It is not the size of the head of a pin, not an object a million times smaller than the head of a pin. It is a point with no dimensions, but super-massive in weight. (I recall from my study of black holes this point is called a singularity.) This singularity is surrounded by pitch blackness.

In the meantime, Hydrojean, who has not consciously deflected herself toward the black hole like Hydro, is nonetheless being drawn relentlessly toward the clutches of the very violent whirlpool of gravity.

"I am doomed to destruction like foolish Hydro." Hydrojean is being overwhelmed by despair and desperation. "Oh, save me, Divine Beings!"

Just at that moment, Hydrojean collides with another particle. Suddenly, she is deflected away from the event horizon of the black hole. Her high velocity allows her to stay out of its annihilating grasp.

"I am saved!" she cries. Relief begins to overcome her anxiety and despair. "Oh, how deeply grateful I am to be delivered from annihilation! Poor Hydro! His pride and independent spirit has brought him to this hellish annihilation. I am profoundly saddened by his demise."

Ruah calls from a distance. "Hydrojean, how happy I am that you have been spared. You have much creative work to do in the future." Ruah recognizes her sadness and grief over Hydro's annihilation, her sense of relief for her salvation from the black hole, and her feelings of deep gratitude. "Hydro brought on his own annihilation, Hydrojean. He closed himself off from all help. His independent spirit, pride, brashness, and curiosity caused his utter destruction and annihilation. How unfortunate!"

"Thank you for your sensitivity and compassion, oh beloved friend, Ruah." Gratitude lights her face with a glorious glow.

"Continue to trust in us, Hydrojean, and in the future. You will not be forsaken! You will be a creator with a remarkable future. Trust and join in the community's empowerment to go forward to more creative activity!" Ruah reassures Hydrojean with these words.

"I do take heart, Ruah, and I rededicate myself to trust in you and in our future creative mission for the development of our beautiful universe. I go forward with great hope." She concludes her reaffirmation.

What amazing events I have just witnessed! I ponder what I have just observed. What is in store for the universe?

Chapter 16
Phenomenal Photie's Tale

In the middle of the afternoon May 10, 5.4 byich, I come upon the radiant Divine Beings conversing in the midst of this gigantic, dark desert of intergalactic space-time. Their dynamic presence illuminates this area of black space-time. I move closer to eavesdrop on their conversation.

"I am still heavyhearted due to Hydro's demise." Sophia's voice conveys her profound sadness. "He had such great potential as a leader in the hydrogen family and in the community of beings. His pride and spirit of independence led to his annihilation in that black hole."

"My beloved Sophie, this is our first big disappointment with our new beings." Elohim seeks to console her. "But it will not be our last. When we gave our community of beings free will, we gave them true freedom of choice. I would much rather create a universe of autonomous individuals who would create than slavish robots who were completely controlled by us."

"Yes, you are right!" Dabar affirms Elohim. "Creators, not slaves to us, has always been our operative principle from the beginning." Dabar recalls the first moment of creation when the first quark and photon came into existence.

The ghostly fairy godmother, Ruah, is hovering over the other three Divine Beings. "My love energy can be more creative when it energizes free beings. Their freedom utilizes this love energy to produce families of atoms imbued with the same liberating spirit." Ruah affirms Dabar and Elohim.

"The three of you are correct. Freedom is an essential ingredient in the creative process. Yet, I still mourn for our dear friend Hydro." Sophia assents to the creative principle enunciated, but affirms her need to grieve over Hydro's demise.

"We all grieve the loss of beloved Hydro, Sophia." Ruah proceeds on to make another point. "But creative history must go on. There will be new and exciting events occurring in the future, I feel confident."

With excitement, Dabar shouts, "I see Photie. He is almost upon us."

"How wonderful!" Sophia catches Dabar's enthusiasm. "We have not talked to him since his birth within the first second of creation."

"I remember that we began a conversation with him at that time." Dabar is recalling the moment. "But he seemed befuddled. He immediately took off without saying one word to us. And he left us with colossal speed."

Elohim greets Photie. "Photie, do you remember us? We are your loving Creators. Sincere greetings to you!" Elohim exudes caring love and sensitivity.

"Resuscitated remembrance, oh Divine Beings, envelopes me." Photie's demeanor is much different from the first moment of his birth. He seems experienced and self-assured.

Elohim has been puzzled by Photie's abrupt departure from their initial meeting. So he asks with gracious sensitivity, "Photie, why did you leave us so impulsively in that first moment of contact?"

"Birth born into an eminent existence, I was filled with frightening fear and weighty wonderment. Overwhelmed by aching anxiety in your ponderous presence, oh Divine Beings, frightened flight was my only option. Or so I reasoned." There is a poetic, alliterative quality to his candid responses.

Sophia smiles, amused by his lyric speech. "Well spoken, dear Photie! Your initial feelings toward us are understandable. But you are so self-assured, self-confident. How did you come to this confident spirit?" All the Divine Beings anxiously await Photie's response.

"Exciting experience has been a testy tutor." Photie gazes with glowing intensity at the Divine Beings. "Frigid fright filled my besieged bosom to the bearable brim as I launched my enchanting existence into the warming womb of the universe. But whirling wonder and accumulating awe soon overwhelmed me and replaced my feverish fear. Wordless wonderment stemmed from my ambling adventure in the dusky darkness of the swelling space-time. Convulsive chaos initially greeted me in the hectic heat and dank density of pregnant, seething space-time. Colossal collisions with other prodigious photons created engaging amber electrons and pretty topaz positrons. Exquisite amber electrons struck pale white protons birthing nifty silver neutrons and naked emerald neutrinos. My personal ecstatic existence seemed jeopardized in this chaotic ecumenical environment." Photie pauses to catch his breath.

"What amazing insights, dear Photie." Sophia is in awe of his vivid descriptions.

"Seducing space-time expanded by gargantuan speedy steps in that primeval, meaningful moment of creation's desirous dawn. I despaired that I would ever depart this creatively topsy-turvy time. Towering temperature subsided and moderated. Crowded, dark density gave way to more luxuriant, spirited spacing. Freed from the imposed imprisonment of thick, demonic density, I was able to escape into liberating, curved, spangled space. But as a puny package of engaging energy reflecting a faint flicker of yellow luminous light, I began to sense my inimitable individuality." Photie concludes his tale with a flourish.

"Photie, did not you feel lonely in your inimitable individuality?" Sophia is thinking of the universe's law of community.

"My filial photon family soon discovered me in this dark, swelling space." Photie's face radiates as he continues his tale. "My fantastic family embraced me with intoxicating, lilting love. Inebriated, I succumbed to their passionate, encompassing embrace. My perilous person was overwhelmed with jocular joy in their caring caress. Frank friendships established, we voyaged out into an almost limitless, unctuous universe with earnest expectation and hardy hope."

"Tell us some more of your adventures." Dabar focuses attention upon Photie.

"Momentous moment was hour three on January 1! Negatively charged amber eligible electron family mated with white positive proton family to create the orange hallowed hydrogen family. No longer could my yellow phantom photon family abruptly disrupt these loving, trusted trysts. The previous torrid temperature had moderated from tempestuous trillions of degrees to mellow millions to a very mild three thousand drafty degrees. Silver noble neutron clan was able to join the white pandering proton family to lead in three stages to the burgeoning birth of golden hectic helium 4 nuclei family." Photie's momentary pause gives rise to soaring enthusiasm within himself. "Feverish alluring activity ceased for me as I journeyed into entrancing, spectacular bent space-time."

"Your observations of early space-time are fascinating." Elohim compliments Photie. "They capture some of the drama of the universe's early creativity." He hesitates. Then he continues. "What else did you observe, Photie?"

"Orange hardy hydrogen and golden heliotropic helium 4 nuclei were attracted to different areas of the luxurious universe." Photie describes their purposeful journey. "Slowly, these gracious, gaseous elements began to create luxurious lumpy islands. Crapulent chaos was giving way to obsequious order."

"Those luxurious lumpy islands of gases we have named galaxies, Photie," Sophia informs him. She is spellbound by phenomenal Photie's verbiage. "What did you see next?"

"Huge, cumulative clouds of gathering gases of halcyon hydrogen and hellish helium nuclei populated these billions of fast-growing, isolated islands." Photie continues his colorful tale. "Passing through an impetuous lumpy island of momentous measurement, I was present to a pregnant process in one part of a monstrous, clannish cloud of gaseous elements. Invisible peerless power drew hardy hydrogen and heliacal helium into a whirling, dervish rotation. Soon, an immense spinning sphere of gases had formed. Within this saffron sphere, heavier helium 4 nuclei fell into the central core and began to build up a tempestuous temperature. Heady hydrogen entered into a three-step process to produce additional hysterical helium 4 nuclei to fuel this cooking core."

When Photie pauses for a moment, Dabar comments. "Photie, the invisible peerless power that drew these gaseous elements into the rotation and caused the heavier helium 4 nuclei to fall into the core is called gravity. Its source is our ecstatic energy of love. This weak force also bonds you to every other being in our beautiful, expanding universe. In addition, those monstrous, clannish clouds are called nebulae."

"Oh, grandiose gravity!" Photie exclaims. "Oh, lilting love energy of you, the Divine Beings! Oh, noble nebulae!" He savors the names and their source. He goes on with his tale. "Hellish helium core heats up to tempestuous temperature. Suddenly, this conflagrant core ignites, releasing numberless numbers of our fantastic photon family to penetrate through the surrounding hydrogen gases. Finally, our photon family lights up the whole of this huge, blue-white, spinning, spectacular sphere, its nebular nursery, and our ubiquitous universe."

"Again, Photie, we have given a name to this huge, blue-white, spinning, spectacular sphere." Sophia names this body. "We have called it a sun. It also has another name, star, because at a distance it is so sparkling. You, Photie, saw your photon family in action helping to light up the universe." After a pause, Sophia continues. "The conflagrant core, as you called it, has a name. It is a fusion furnace."

"Oh, scintillating sun! Oh, beautiful body of light! Oh, furious fusion furnace!" Photie delights in pronouncing the names of these new creations in which the photon family has had such a prominent part. He continues his story. "Within a hoary hour, as I continued my jubilant journey, millions of other fusion furnaces were igniting in each of the millions upon millions of contagious cloud nebulae in millions of gay galaxies, creating millions of ceremonious cities of light. Our phenomenal family of photons is the source of all these luminous lights."

"But what of your recent journeying?" Elohim stimulates Photie to recount his most recent adventures. "Tell us what you observed."

"Just recently, I passed by an oblong, gracious galaxy. Most numerous of all the sun cities, oblong gallant galaxies are usually devoid of stellar nestling nurseries. The paucity of genial gases and dreary dust limit its ability to become pleasingly pregnant with new blue-white sumptuous suns." Photie goes on to describe this particular galaxy. "Yet, spanning these immense ambient areas of space-time, I saw many, many stunning suns in its dazzling disk. This city of spectacular suns gleamed and sparkled as I sped by." "

"Photie, these oblong, gracious galaxies are called elliptical galaxies." Dabar again names this body. Then Dabar questions him. "What was the most amazing sight you observed, Photie?"

"Early in my jaunty journey, I observed a huge, red, swollen sun suddenly begin a colossal collapse upon its fiercely burning carroty core. Gracious gravity tugged the luscious layers of elegant gaseous elements downward, creating colossal pressure. Massive decadent density, red-hot heat, and enormous energy ensued. Pressure-packed, millions of degrees temperature, the compact core manufactured new, amazing atom families, light, medium, and heavy. Elemental benevolent birthing process ended. But this disintegrating stunning sun was still pregnant with searing surprises and cosmic creativity. The imploding rabid red giant shuttered briefly, shedding its element-rich lofty layers left orphaned by its collapsing contraction."

Photie is excited as he recounts the last gasp of the disintegrating red giant sun. "Like a massive venting volcano, this sad, dying sun exploded, blowing off its enormous exterior. Stupendous hectic heat generated by this hyperblast birthed even heavier ardent atomic families. This expansive explosion and its accompanying shock waves shoot these aerodynamic atom families into the ponderous path of wispy, greedy, nifty nebulae, which absorbed them. This entire nearby sun-studded elliptical galaxy I had visited recently felt this frenzied alacritous assault."

"Then what happened, Photie?" Sophia eagerly awaits his account.

"My gluttonous gaze was directed back toward the ravaged sun's swiveling crumbling core that had survived this bombastic blowout. Gargantuan gravity crushed the remaining resident atom families, creating naked silver neutral neutrons. Shoulder to shoulder, these neutered neutrons were stuffed with no smidgen of space between them. Precocious pressure, devilish density, and gentle gravity pushed this rounded, crushed core of this stricken star into a more and more rapacious rotation. Very powerful magical magnetic field was created. Powered by this massive magnetic field, energized rhythmic radiation shot out of the crushed core in opposite directions at almost sublime speed. Two bewitched beams of rapturous radiation emanating from opposite sides of this crushed core of noble neutrons pulsated lucent light like a rotating, beckoning beacon. Magnificent was the stunning sight of this rapidly blinking, dead, stuttering sun." Photie concludes with a note of satisfaction about his vivid description.

"We need to give this new sun-being a name." Dabar advises.

Elohim responds to the request. "Since the crushed core of this sun remnant is made up of only silver neutrons, let us call it a neutron sun or star. Its twin beams of radiation we name a pulsar. It will shine like a beckoning beacon in the sky to the rest of the community of beings in our universe."

"An appropriate name, my dear Elohim." Sophia commends him.

Photie has been quiet during the naming. He approves of the sun's name. There is a pause in the conversation. Photie has been musing about the future. He finally says, "Proud profundity is transpiring in this sun-flickering, whimsical womb of the universe. Up to the present time, I have been mainly an overawed observer of entrancing events. I have an intriguing intuition that I am about to enter a more active and tantalizing creative time in my eminent existence. I cannot wait to become a compassionate creator, oh beloved Beings. I feel cuddly confident about my flashy future and that of the unctuous universe."

Ruah finally enters the conversation. "You have captured some of the drama of the major events of the creative history of our expanding universe. Whether you realize it or not, you and your photon family members have already played a significant role in the cocreation of the atom community in the first few seconds of creation." Ruah then comments on Photie's intriguing intuition. "You are right, Photie! The future is pregnant with possibilities. And you will play a significant role in the universe's unfolding creative history."

"I trust you, Rousing Ruah. Also, you, Elohim, Sophia, and Dabar." Photie is bursting with trust and hope.

Ruah reassures him. "I am always close at hand, Photie. I will energize your photon family and all the other atom families as you create an expectant universe."

Photie concludes his remarks. "My peerless person is open and poised for unimaginable crowning creativity in union with you, my beloved Beings."

Chapter 17
Beacons in a Vacuous Ocean
of Space-time

I am now alone, and my vision transports me on a spectacular, arced space flight. The date is August 1, 8.7 byich. I look out upon a gigantic ocean of almost vacuous blackness. This ocean is like the melding of millions upon millions of Pacific Oceans into one.

In all directions within the black universe bubble, I see vast cities of lights. These galaxies contain hundreds of millions of points of light, blue-white, yellow, orange, and red. My optic wizardry, due to my powers as a visionary, allows me to observe how all of these cities of light and all of their individual lights are constantly orbiting around the central core of their disks. The light-packed cities, the galaxies, are gyrating in relationship to each other. Yet, rarely, as I previously mentioned, do any suns collide within each galaxy or any galaxies smash into each other. The vast distances between suns within a galaxy and between the galaxies rarely allow this contact. Since all these communities of suns are moving in concert with each other, it appears that there is little or no motion. (This situation is similar to my life on earth, whereby my planet Earth is rotating on its axis and traveling at about sixty-seven thousand miles an hour orbiting our Sun, but I do not feel or experience this extremely rapid movement.)

Even though there are these billions of galaxies containing hundreds of billions, and even trillions, of stars, their accumulated light does not erase the blackness of this limitless ocean of intergalactic space-time. Yet, these points of light act as small beacons of hope in the dark desolation of intergalactic space. Some of the suns appear faint and pinkish due to intervening clouds of intrastellar dust, which scatters the blue light more than the red. I am seeing these faint, pinkish suns as through a dense fog of nebular dust. Other stars are distorted in brightness and color by being obscured behind swirling, whirling rose, azure, aqua, and golden yellow nebulae. If it is in closer proximity to me, the sun is brighter.

This ocean of black space is not devoid of matter. I again pass by some primeval clouds of hydrogen and helium gas drifting in this vast, oceanic expanse. My powers as a visionary allow my eyes to possess the ability to magnify and identify minute traces of atoms of various kinds: hydrogen, oxygen, carbon, and nitrogen. Each atom has his or her own story of cocreative activity. They have been blown out into intergalactic space-time by supernova explosions, rare collisions of galaxies, or other strong forces. Under ultraviolet radiation bombardment and other advantageous conditions, some varieties of atoms meet and create primitive compounds or molecules of water, ammonia, formaldehyde, and methane. Atoms and molecules attach themselves to tiny grains of dust as they journey through the universe. Intergalactic space-time is much cooler with very low density. On the other hand, in intragalactic space-time, the area between suns in galaxies is hotter and denser.

From August 1 to August 13, 8.7 to 9.2 byich, I traverse through this gigantic, sparsely populated, ebony ocean. On August 10, I approach a supercluster of galaxies spread over about 210 million light years. First, I observe a local cluster consisting of several elliptical galaxies, with a lesser number of spiral galaxies, and fewer irregular-shaped galaxies. Among all of these are some dwarf galaxies. Each one contains hundreds of millions of suns.

I explore the interior of a nearby oval-shaped galaxy. Within its boundaries are billions of yellow-colored suns. This color indicates that they are mature, stable suns, having been shining for 6 to 7 billion years. They began to be birthed in this galaxy's stellar nurseries about January 5, 200 myich. There are a lesser number of orange stars and still fewer red stars. There are remnants of red stars that had gone supernovae and are now neutron stars. A few of these latter stars have become pulsars, wildly spinning neutron stars pulsating their beaconlike light to the rest of the universe.

Suddenly, I am stunned. A massive red star blows itself apart. The brilliant cloud of exploding gases and dust lights up the galaxy and its neighboring sun cities in vivid colors of red, saffron, yellow, pink, violet, and blue. The enormous energy of the blast sends tumultuous shock waves and photonic radiation through this intensely glowing city of suns. The gases and dust are blown into intragalactic space-time, although some of them escape the galaxy. I continue to look back hourly to watch the red, yellow, blue, and violet swirling clouds of gases expand into the far reaches of intragalactic space-time.

On August 17, 9.4 byich, I visit a strange new galaxy. It reminds me somewhat of a spiral galaxy. It has a round, disklike nucleus. The latter contains billions of suns. Again, the stable yellow stars predominate. In its center, I notice a whirlpool of activity. It is a huge black hole. At its event horizon, I see gas, dust, and part of suns being dragged ominously by its fearsome gravity over the edge and into its incredible tornadolike whirlwind. As happened to Hydro earlier, the atoms are catapulted into the insatiably hungry black hole. There they are elongated, torn into subatomic particles, and compressed into a point of massive density or singularity. Again, I am appalled and overwhelmed by the voraciousness of this awesome natural phenomenon.

Above the galaxy's thick, platter-shaped nucleus and core, I discover a vast halo or canopy. Under this dome are clustered millions of older suns. They orbit the core. When I emerge from this dome, I notice that there are two gigantic arms growing out from its central disk in opposite directions. The distance from the far edge of one arm to the far edge of the other arm covers fifty thousand light years. Along these straight arms are strung bright blue-white suns, young and energetic. I am privileged to observe a number of nascent suns spin out of their multicolored clouds and suddenly ignite their fusion furnaces to radiate their newborn light to an expectant universe.

At the far end of each long bar emerges one spiraling arm. Again, these arms are filled with billions of suns, many of them young blue-white stars. Each of these curving arms is filled with enormous cottony clouds of yellow, red, pink, violet, and saffron-colored gas and dust. I again watch one nebula begin the birthing process of new suns. Finally, a new baby sun is born and continues its circling around the core of this bar-shaped galaxy. (Again, I recall from my science class that this type of galaxy is called a barred spiral.)

On August 23, 9.6 byich, I emerge from this supercluster of galaxies. I have mapped out a bowed journey that will hopefully bring me into contact with my atom family friends and the Divine Beings in a far-off local cluster in this dark, expanding universe.

I am again in open intergalactic space-time. Again, I am overwhelmed with awe and wonder by the immensity of this ocean of space. Following a curved path, I float in the vacuum of space-time like the astronauts we see floating in the weightlessness of their space vehicle cabin.

I traverse this vast, spatial ocean and enter this local galaxy. I am entranced by the majesty of several spiral galaxies I observe. To the flowing strains of a Strauss waltz played by the subatomic strings and loops of strings, the spiral disk and its majestic arms slowly rotate around its center core. Its arms are slowly opening a rhythmic motion, like entranced dancers, continuously circling as one to a waltz tempo. They glide inexorably in their orbit.

On August 27, 9.8 byich, I come upon another local cluster. A dramatic event is transpiring. Two spiral galaxies are on a collision course. By evening of August 30, 9.9 byich, the outer edges of their arms are beginning to come in proximity to each other. Gravity is gradually pulling these sun cities into a catastrophic battle. Already, the millions upon millions of suns on the outer edges of the arms are being violently disturbed by gravitational forces from this invading galaxy. A few hours later, these arms are violently colliding into each other. Millions of suns are thrown off course, crashing into each other. Suns explode, sending their remnants of gas and dust into the intragalactic space of this local cluster.

By the afternoon of August 31, 9.95 byich, both spiral galaxies have suffered major distortion to their curved, spiraling arms. But each spiral endures this glancing blow. Gradually, as the day goes on, I notice that the collided galaxies are moving away from each other, bearing the huge scars of this titanic crash.

In this local group, I observe some very irregularly shaped galaxies. Each of these immense star cities contains billions of stars of differing age and color.

* * *

It is the morning of September 1, 10 byich. My August odyssey has been a momentous one. My explorations have given me a much better understanding of the stupendous dimensions of this black, swelling womb of the universe.

As I curve my way through a supercluster of galaxies, I am approaching another local cluster. I notice that Elohim, Sophia, and Dabar are present with Ruah. I see the motley companions, Quarkie, Quarko, Quarkoff, Hydrojean, Nitro, Phosie, and Photie, moving along with them.

I recall Photie's comment back in May: "I have an intriguing intuition that I am about to enter a more active and creative tempestuous time in my effervescent existence." And Ruah's response to him was: "You are right, Photie. You will play a significant part in the universe's unfolding creative history."

I have a premonition that the atom families and the Divine Beings are very close to the time of immense creativity. Where will their cooperation and creative activity take them? I cannot wait to discover the next leap forward in the history of this dynamic universe!

Chapter 18
The Divine Beings Give the Atom Characters a Guided Tour of a Spiral Galaxy

It is September 1, 10 byich. I am about 160,000 light years from a magnificent spiral galaxy. Its clockwise-rotating central disk and core are moving effortlessly in concert with a tranquil melody suffused with delicacy of tone. The musical accompaniment is like a tonal pictorial painting of Delius, sensitive, poetic, and sensuous, depicting the grandeur and beauty of this serene scene. The core is gently dragging its arms, which flow delicately from several areas on the edge of the disk. I observe that its long, curving arms are gigantic stellar nurseries. From time to time, I watch a neophyte sun ignite in various section of each arm. Bathed in the tranquil musical accompaniment of the subatomic strings and loops of strings, I never lose a sense of awe and wonder at the time of stellar birth.

I find the Divine Beings and their atom friends. They are conversing as they journey together.

As I arrive, Dabar is saying, "We are just passing a gorgeous nebula. Do you notice how your hydrogen and helium 4 families are working together to birth new stars?"

"I like the gauzelike appearance of this huge cloud." Quarkie takes note of the dreamy, wispy nature of this huge nebula. "Its rose color is so royal."

"Behind those billowy rose clouds are many rotating clouds of our friends in the hydrogen and helium 4 families, as you mentioned, Dabar." Quarkoff remembers his own experience. "It takes great effort and much suffering to work in those intensely hot conditions to ignite the fusion furnace and produce sunlight. And then to come together in intimacy with two other helium 4 nuclei to birth a new atomic family, the carbon family." Quarkoff is profoundly reliving this experience through his storytelling.

There is an amused expression on Quarko's face as he says, "Yes, I felt like a quick-change artist again. One moment, I was a quark-proton that was part of helium 4 nuclei family; then I was a member of the carbon family. I have lost two identities already in my 10 billion years of existence."

Elohim smiles when Quarko begins his humorous comment. Then the former's countenance becomes more serious as he says, "You may need to continue to cultivate your talent as a quick-change artist in the future."

"Then I'll try not to take my present identity too seriously." Quarko deadpans his comment. Everyone laughs.

"You need to keep that sense of humor, my dear Quarko," Sophia affirms. Then she continues. "I have an intuition that it will not be long before you will be involved in some momentous new enterprise. I have a premonition that this majestic spiral galaxy in front of us may be the location of new creativity."

The group is passing by a very sizable group of nebulae similar to those they had just passed in the spiral galaxy. Elohim says, "Let me give you a tour of the galaxy and its neighbors. You notice immediately that the two nebulae we just passed are orbiting around the core of this spiral galaxy. Observe the other twenty or so galaxies of various kinds that are part of this cluster. They, too, are related to you by gravity that holds them together as a family group."

"I notice that the long arms of this spiral galaxy rotate around a disk-shaped center. Tell us something about how the galaxy operates, Elohim." Hydrojean is inquisitive.

"As you notice, the round center disk and all its arms make up a plane, something like a thick disk." Elohim begins his description. "This flattened platter of the galaxy is about one hundred thousand light years in diameter reaching from the arms on one side to the arms on the other sides. The thickness over the round center is about ten thousand light years. These are immense distances. Within this disk, not counting its arms, there are arrayed hundreds of billions of stars."

"Elohim, tell us something about the round center of the galaxy." Nitro, being a member of one of the youngest atom families, wants to learn all that he can about this galaxy that might impact his life shortly.

"The round center is called a nucleus. It is about sixteen thousand light years in diameter. That's about 15 percent the size of the whole spiral disk." Elohim is going into detail. "The nucleus contains millions of older, mature yellow suns. Some are the red suns, possible supernovae in the future. These are bigger stars that live much shorter sun lives. A few are neutron suns or stars, signaling to the sun community through their flashing beacons of pulsating light."

"I notice something unusual in the center of the nucleus." Hydrojean is interested.

"Its center, called the core, will recall sad memories to you, my dear Hydrojean." Elohim's voice takes on a grave tone. "That is a black hole like the one that annihilated our friend, Hydro. The black hole almost grabbed you, Hydrojean, into its tornadolike whirlwinds."

"I was deflected from being sucked into that deadly whirlpool by striking another particle of matter. How fortunate I am to still be here." Hydrojean is expressing her deep gratitude for being saved from obliteration. But Hydro is brought back to memory. "Poor Hydro! I tried hard to stop him from following his foolish, independent curiosity. But he would not listen. Once he went over the edge of the event horizon and felt the massive power of the surging gravity, I am sure he understood his tragic mistake. But we are all survivors, Elohim. We trust you. We want to be a part of the community of creators, Sophie, Dabar, and Ruah."

"We are grateful too, my dear Hydrojean, that you are still with us," Sophia continues. "We too feel the loss of our beloved Hydro. He was indeed stubborn. And this led to his extinction." Sophia looks to Elohim and says, "But continue your enlightening tour log."

"Notice the thin domelike structure over the entire disk. It is one hundred thousand light years in diameter and is called the halo." Elohim says more about the halo. "Underneath the halo are clusters of old yellow stars. Notice also that when you look at this spiral galaxy from the side, rather than from above, the halo gives it the domelike look."

Phosie has been listening and correlating Elohim's information about this galaxy with her visual observations. She directs a question to Elohim. "What is that very, very large, mysterious-looking canopy overarching the halo, disk, nucleus, and black hole?"

"That large canopy is called a corona. It is very, very massive in its expanse, about four hundred thousand light years in diameter, to be precise."

Dabar intrudes and then explains. "In the first moments of creation, there was created a very dark matter and dark energy. It has been in the universe since the beginning, hovering over all the galaxies. It has exerted additional pressure and energy upon all the matter in the galaxy, and it helped all of you to begin the first ignition of a fusion furnace and all of the ignitions since that time. The pressure of dark matter and dark energy helped to energize the red stars to go supernovae and create all of the atom families of which you are a part. Beneath the corona is where dark matter dwells in all the galaxies."

"Of what does dark matter and dark energy consist?" Phosie is inquisitive.

"That's the mysterious part of this matter and energy, Phosie. You will find out someday. In the meantime, I would just like to let you know that dark matter and dark energy constitutes about 90 percent of all the mass in this magnificent universe. It has a significant role to play in the creative activity of the universe." Elohim finishes this part of his lecture on the spiral galaxy.

"What name do you give to this magnificent spiral galaxy, Elohim?" Quarkoff asks.

"I name it the Milky Way galaxy." Elohim is quick to respond. "As you look back from distant space-time, this galaxy exudes a hazy white color. Hence it's name, Milky Way."

Quarkie has been attentive but quiet for a long time. She finally speaks up. "Elohim and Dabar, you haven't told us about the immense, curving arms that emanate in various places form the nucleus of the Milky Way galaxy. Tell us about its arms."

Elohim begins his response. "Let me take you along one of the particularly gargantuan arms filled with many blue-white stars and a few interspersed red stars in order to describe the operation of these arms."

Sophia has a significant prophecy to make. "A short time ago, you might remember, I said that I had had an intuition that a significant momentous event was soon to happen." She pauses for a short time. Suspense begins to build with the atom family members. Finally, Sophia continues. "This is my intuition. It is in this spiral arm of the Milky Way that you may all be engaged in a profound adventure."

"Tell us more about this event, Sophia." Quarkie is insistent.

"Be patient, Quarkie. We must finish this tour through these intriguing spiral arms to find out what destiny has in store for all of you." Sophia is adamant. Everyone must be patient.

Chapter 19
A Journey toward a Special Destiny

Is this a journey toward a special destiny? Is this spiral arm the locale for this special leap forward in creative history?

Members of the atom families—Quarkie, Quarko, Quarkoff, Hydrojean, Nitro, and Phosie—begin to explore one of the immense spiral arms that has spun off the nucleus of this Milky Way galaxy. They are in intragalactic space. It is September 5, 10.2 byich.

Even though the entire spiral galaxy is rotating majestically around its center core, gravity is keeping each of the billions of suns in this vast sun-studded arm in its proper relationship. In a far-off corner of this massive spiral galaxy's arm, they see occasionally two or more suns migrate by gravitation into the path of another, more massive star. They begin to interact and affect each other's normal orbit and way of life.

The ebony space between the suns is like hundreds of thousands of Pacific Oceans set side by side. Their path traverses past white dwarf suns, which had lived billions of years of sun life and now are a tiny shell of their former existence. They observe brown dwarf suns. These are stunted suns that never reached the critical temperature of 18 million degrees to start their nuclear fusion furnaces. Brown dwarf suns look like hot, fizzled embers in relation to real suns. They pass middle-aged yellow suns. Finally, the group observes a massive, aged red sun, which expands, implodes, and explodes, leaving this small white shell, a white dwarf, as a reminder of its previous existence.

Most of this vast space in this very long spiral arm is made up of millions upon millions of massive nebulae populated by young blue-white suns. Occasionally, there are middle-aged yellow suns. But this is chiefly the realm of youthful suns.

These kaleidoscopic nebulae are the stellar nurseries for new baby suns. In each colorful nebula, the density of gases and dust varies widely in different sections of this massive cloud. They are massive, two thousand light years in length and one thousand light years in width and depth.

Where the density of dust and gases is high, gravity causes them to begin to coagulate and rotate. Galactic winds from nearby supernovae or other sources further agitate this knot of dust and gases. The lighter hydrogen follows the heavier helium 4 nuclei as they descend into the core of this wispy, saffron, tightening knot of cloud. Further pressure exerted by gravity, the increasing density within this gaseous knot, along with the pressure exerted by dark matter and dark energy, increases the internal temperature in its helium 4 core in excess of 18 million degrees. At this point, this nuclear furnace turns on, releasing a mushroom of radiation, which radiates and begins to illumine this ignited knot of cloud. A new blue-white spherical star is born to illuminate this spiral galaxy.

In this massive nebula, tens of thousands of knots of clouds are acting similarly and peopling this section of the spiral arm with bright, new illuminated inhabitants.

* * *

Very early on September 10, 10.4 byich, as I accompany, from a short distance, the motley group of atom friends through this spiral arm, we pass by a huge, still-expanding red sun. Together, we watch its ballooning process. It is a magnificent, awesome sight.

My atom friends again encounter the Divine Beings. Elohim, Sophia, Dabar, and Ruah are majestic and sublime in appearance. Their mutual love illuminates and unites their persons.

"Greetings, our dear atom friends!" Sophia begins the conversation. "We last saw you on September 1, 10 byich. We hope that your journey through this arm has been enlightening." She adds, "No pun intended."

"You gave us a grand tour of the nucleus, black hole, halo, and corona of this grand spiral galaxy." Nitro is the first atom member to speak. He has been standoffish and shy in the past. "Elohim, your explanation helped to ready us for the spectacular sights we have experienced these past ten days."

"Yes, we have been treated to the extraordinary work being done by my hydrogen family, along with the helium 4 family." Hydrojean is proud of her family.

"Remember, Hydrojean, I was in the helium 4 nuclei family at one time." Quarko remembers his origins. "Although I had a quick change of identity in that imploding red giant when it went supernova (please admire my artistry as a quick-change artist!). I did at one time belong to the helium 4 family before joining the carbon family. I am proud of my relationship to the helium 4 family." Quarko's face exudes pride but is tainted with a mirthful smirk. "I guess I am an unsettled, quirky quark ready for a far-out future."

"Quarko, your 'far-out future' may not be too far distant." Quarkie counters through a broad smile. "Sophia intuited ten days ago that a momentous event possibly was to soon happen and Elohim told us that it would occur in this very spiral arm where we are right now."

"You still need to be patient, Quarkie." Sophia reminds her of what she had said at the conclusion of their September 1 meeting.

It is at that very moment that Photie comes into sight. "My dear atom friends, I want to introduce to you a member of the photon family." Dabar makes the sighting and alerts everyone. "He is just coming into sight. He is yellow in color and not of the atom families. He is a weightless package of pure energy."

Photie is just coming into sight as a dot of yellow light. "Photie, so good to see you again." Dabar extends the greeting to him. "This is an opportune time. All our friends in the atom families are present. They have never met you."

Quarkie looks carefully at Photie. "I remember you. During my very early existence as a proton, when the temperature was extremely high, as was the density, you and your photon family would continually disrupt us just as we were about to mate with a free amber electron. All of you were big pests." She smiles as she concludes her remarks.

"You continually broke up our torrid love affair with the gorgeous amber electrons." Quarko breaks into the conversation. "You were a naughty nuisance packing quite a wondrous wallop. You stopped us early from birthing the hydrogen family."

Photie is taken aback by Quarko's remarks, not knowing his witty, biting personality. Momentarily, he reverts to his shyness. He remains quiet.

Dabar reads Photie's reaction. "Don't be heedful of Quarko, Photie. He is our resident comic and cutup."

"You know how to really put down a friend, oh dashing, divine Dabar." Quarko looks hurt by Dabar's comment. Suddenly, his countenance lights up and he smiles broadly. "I accept your kind compliment. I guess that is a good definition of my personality." He pauses for a moment and then shouts, "Thank you! Thank you kindly."

Photie lightens up and smiles. "Ferocious fear enveloped me when you first began to address me, quipster Quarko. But delightful Dabar has eased my haunted heart with his translucent, classy clarification."

Dabar is again amused by Photie's lyrical language. "You can see that Photie is our resident lyricist and poet. So we have a quipster and a poet among us."

"Your daunting definition of me, Dabar, pleases my soaring soul." Photie is happy with Dabar's description of his person.

Dabar proceeds with his introductions of Photie to his atom family friends.

"Photie, about the third hour on January 1, when the temperature cooled down drastically and the density lessened, we could resume our red-hot romances with the amber electrons." Quarkoff is recalling those early hours of creative history. "Your family's once-wicked wallops lost their power to disrupt our mating. Then we lost our identity as pure quarks, and a white photon and became part of the orange hydrogen family. From there, we could mate with silver neutrons and become members of the golden helium 4 nuclei family."

"In the fierce fire of those early horrendous hours of hope-filled history, we fancy photons were reluctant crippled captives of daunting density that drove us to be pesky pests toward proud protons." Photie defends his family and himself for their seemingly errant behavior.

"When the heat descended to a moderate range and the density decreased greatly, our roaring romance with the beautiful amber electrons became timidly tepid." Quarko comments cutely in imitation of Photie's poetic style. "Our marriage was a success. We became creators, oh devoted Divine Beings. Our web of creation tightened." Quarko was serious for once.

"The time may draw near soon for a very momentous event in which all of you may share." Elohim pauses and reminds them. "It was Sophia who intuited that this fateful moment may be near and that this special place is the site of the momentous event." Elohim brings the group back to the task at hand. "How wonderful that all of you, our dear atom and photon friends, are present at this location to possibly take the next profound step in fulfilling the destiny of our beloved universe." Elohim looks intently at his atom friends for a brief moment.

Then Elohim continues to set the scene. "You are about forty thousand light years out on this spiral arm within this wondrous galaxy. You are presently entering deeply into this nearby gaseous nebula, the birthplace of new stars. You will soon encounter a dense knot of gas and dust."

Quarkie becomes excited. "We have experienced already being partners in the birth of new suns. We have been present to the tornado like turbulence of an imploding and exploding red sun. Will this part of our mutual journey be a repeat of these past experiences, Elohim?"

"Elohim, let me respond to Quarkie." With calm, Sophie intrudes herself into the discussion of the future. "Your adventure may involve the living through of some of your previous experiences but entry into events altogether new." Sophie pauses. Everyone is attentive, waiting for additional information. "Ruah, as always, is with you at very ordinary and special times to energize you in your mission on behalf of the web of creation and your incredible universe." Again, there is a momentary pause. The members of the atom and photon families are focused and in readiness. Sophia resumes with a question. "Are you ready to be creators once again?"

"We are indeed ready." Hydrojean picks up the unanimity of the group and signals their readiness.

"We trust you, oh Divine Beings!" Quarkie, with her usual enthusiasm, is pledging her deep trust and faith.

"I guess I will have to use my quick-change artistry again." Quarko adds some levity. "Yes, I am ready too!"

"We do not worry about you, Quarko. You're so pliable." Sophia smiles her comment.

"Quarko, the acquiescent clown. That's me, oh Divine Sophie!" Quarko has to get in the last word.

Nitro prophesies, "I too have the feeling that a spectacular event is about to happen here in this nebula, as you have intuited, Sophia."

"All is in readiness." Elohim is trying to reassure them as a response to their pledge of allegiance to this new mission. "Ruah is overshadowing this location. Can you feel her energy of love?"

"I do! I do!" Photie cries with enthusiasm and surprisingly with few words.

"I am ready to be a creator, Elohim, Sophia, Dabar, and Ruah." Nitro pledges his loyalty and trust.

"Definite, divine destiny awaits us." Photie is calm and expectant.

"There may be a provoking detail that needs to occur." Sophia adds the last ingredient to her intuitive scenario.

"What is it?" Hydrojean blurts out. She is anxious to hear what this provoking detail may be.

Sophia ends with a call for readiness. "You must wait in patient expectation." I do not sense any fear or anxiety in any of the quark, atom, or photon family members, but expectation, hope, and trust.

What is this last ingredient to this scenario? What will this momentous event be? I wait, filled with impatience.

Chapter 20
An Exploding Supernova—
The Provoking Event

A tornadolike shock wave from a nearby exploding supernova tears into a massive, colorful, wispy nebula near the point where there is a rotating spherical knot of dust and gases in the early stage of becoming a neophyte sun. At the point of the explosion, a brilliant, crimson-white flash illuminates the nearby space-time, extending several light years in all directions. Energized gases and dust penetrate this whirling clump of bright orange and yellow gases, the proto-sun. This huge nebula assimilates much of the energy and debris shot out from this exploding supernova—hydrogen, helium, oxygen, carbon, phosphorus, nitrogen, silicon, iron, and virtually all the elements up to and including heavy, unstable uranium. These atoms were born in the gigantic temperature, high density, and pressure at the core of the now-exploded supernova. This blast, accompanied by intense photonic radiation and other advantageous conditions, allows some of these elements to form molecules of methane, ammonia, formaldehyde, ionized hydrogen, and water. Many of these atoms and molecules are piggybacked onto very small dust particles. My earthly eyes are unable to see these minute clumped grains of matter. But my eyes, as a visionary, magnify these microscopic-sized molecules and dust particles, so I can see them plainly.

It is early afternoon on September 10, 10.4 byich. I realize that this red star, which has gone supernova, was the one the Divine Beings and the atom characters had passed late on September 9.

My atom friends, who are located about 100 million miles from this proto-sun, gaze with awe and wonder at this magnificent, powerful event.

The suddenness and horrific force of the supernova's detonation stuns me momentarily. At first, I fail to notice the awesome presence of Ruah, the divine, ghostly godmother. She is overarching this massive nebula, and particularly, the area of the rotating, spherical neophyte sun. Her creative energy of love peacefully permeates this newly shocked, devastated part of the nebula. The latter is in chaos, but I intuit that this helter-skelter condition will soon give way to creativity and order.

My visionary gift again allows me to move from the vicinity of the rotating sun-in-the-making instantaneously to the neighborhood of my atom friends. There I observe large, energized meteorites, slim, graceful comets, and gray asteroids of varying sizes, in addition to sizable traces of galactic dust and gases, traveling at high speeds. They had been rocketed into space-time and intragalactic space by the exploding supernova. Gravity is drawing them into a special relationship with this sun-in-formation.

I move back into the vicinity of the highly agitated proto-sun. As I have witnessed numerous times in other nebular star nurseries, the hydrogen atoms are pressing upon the heavier helium 4 nuclei and sink into its core. The temperature at its core is approaching 12 million degrees and continuing to rise. It is still far from the 18 million degrees needed to start nuclear fusion and ignition. I notice that other proto-suns are being formed in this same huge nebula but five or more light years away from this fetal star. All of this star-birthing has begun due to the explosion of the supernova.

In the area where my atom friends are viewing this awesome event, I notice that some of the massive debris in the surrounding space-time is being attracted by gravity into communities of matter. Initially, I observe one mountain-sized asteroid crash into several other hill-sized pieces. They crash at tremendous velocities, liquefying on impact. This releases vast amounts of energy, raises the temperature enormously, and ignites all the matter, forming red, molten plasma. This fire is fueled by the high-velocity impact of the debris and the decaying of radioactive uranium, thorium, and potassium delivered by the impacting asteroids.

Additional large and small meteorites and comets with long tails are attracted by gravity and crash into this growing mass of molten plasma. Its size continues to grow from the constant bombardment of galactic debris. This massive influx of primeval debris enriches this molten mass with many families of atoms and other chemical compounds created by the atom families. The surrounding atmosphere becomes the home of rich carbon- and hydrogen-based molecules, methane, water, ammonia, and hydrogen sulfide. Out of this chaos arises abundant creativity and some order.

* * *

It is very early on September 11, still about 10.4 byich. Fatefully, all of the atom friends, including Photie, survive the supernova's violent explosion and end up in orbit around this particular molten fireball. (I'll let Photie tell his story later.) They are joined by the Divine Beings.

Hydrojean begins the conversation. "Sophia, is this supernova explosion and its consequences 'the provoking detail' that you said must happen to begin this momentous event in which we will be sharers and cocreators?"

"I did predict a provoking event, but I didn't know exactly what kind of event it would be." Sophia begins to respond to Hydrojean's question. She continues. "True, this supernova did provoke the rotating knot of gases to become a spherical sun-in-the-making. Right now, it is still a proto-sun, a fetal star." Sophie affirms Hydrojean's observation. "But the exploding red star provoked another momentous event. It is giving birth to a new being, this nearby molten fireball. It will wander in orbit around this proto-star, which is destined to become a full-fledged sun. Gravity has captured this molten fireball and you and has destined you to be a part of this wandering, fiery mass during its journey into the future."

Elohim speaks up. "Let me give a name to this wanderer. We will call it a planet, a being that is connected intimately by gravity to its budding sun."

"But this planet will be one of several planets that will be drawn by gravity to this sun-in-formation and circle around it in different orbits." Dabar gives them some additional information about what other beings have been created from the supernova's provoking explosion. "Look in various directions. You can see nine distinct planets-in-formation like this planet of which you are a part. All of these planets are bound in community by that special love force, gravity. From the beginning of space-time, I have told you that gravity binds you to everything that exists in this vast universe." Dabar pauses and then continues. "Someday, we will take you on a journey to visit these other planets." The atom characters look into space-time and see indistinctly these other planets-in-formation. He then reminds Elohim, "We must give a special name to this newly forming planet to distinguish it from the other planets orbiting this sun-to-be, Elohim."

Elohim does not hesitate. "We will call it Earth because it will be the home for all of you, our friends."

"A new home!" Quarkie reacts to Elohim's naming of their new home. "Does this mean that we won't be continuing to wander through this vast universe anymore? We will be established in a special planet-home, the Earth?"

"That's right, Quarkie!" Sophia responds. "But the fact that you will be limited to this place will in no way deter your creativity. In fact, your creative work will accelerate in marvelous ways, I deeply believe."

"A provoking event?" Quarko inquires but then comments on the defining event. I am not sure where this nimble-tongued comic will take his inquiry and comment. "In our proton, I am jammed in between Quarkie and Quarkoff. The shock wave of the supernova crushed us together. Talk about an intimate relationship; we were more than bosom buddies. I thought our trio would end up like a scrambled omelet. Being a part of a carbon atom, this violence and enormous temperature threw our six negatively charged electrons off-line. Talk about being positively charged! I became a manic, quirky, quaking quark." Quarko grins. "Soon, the electrons got back online, and our lives returned to quarking normality." He pauses. "So much for an uneventful provoking event."

"But you did survive in one piece—one traumatized, but surviving, carbon atom." Sophia hides her amusement behind a stoic gaze.

Quarko gives a quick response. "We quarks are survivors, Sophia! But classy, cagey, creative survivors!" He quickly adds with boasting and gusto, "Give me the next challenge. This quick-change artist is ready to perform some additional magic."

"Challenges you will have, friend Quarko." Dabar resonates with the comical quark. "And magic! Your crusty creativity will bring forth magic about which you have never, ever dreamed."

"Quarko, we have just assimilated this challenge from the supernova's explosion." Quarkoff comments on his boasting attitude. "Let us have some time of peace and serenity. So lay off this talk about the next challenge, Quarko."

111

Phosie interrupts this interchange with a question. She has a worried tone to her voice. "Has anyone seen Photie? I was right next to him when the shock wave struck. But he has disappeared. I am worried about him."

Everyone pauses and looks around with concern. Photie is nowhere in sight. Phosie's deep concern passes to the whole group of atom friends.

Suddenly, the solemn quiet of the scene is broken by a gentle but familiar voice. There is a release of tension and sense of profound relief.

"Vacuous violence of exploding, rampaging red sun transformed my fragile frame into a sparkling new idyllic identity." Yes, this is certainly Photie but dressed in a novel disguise. He continues in his sonorous, noble style. "This shocking, whipping wave tossed me pell-mell into the pithy path of a neutral, orange, homey hydrogen atom. Being only a power-charged package of elementary energy, my crunching charge invaded the royal realm of a spinning elastic electron. My savage surge struck it solidly. This negative pliable particle was released from its orderly orbit to frivolous freedom in galactic seamless space. My peerless person now positively charges this formerly highfalutin hydrogen atom."

"Just meeting you, Photie, always brought me a positive charge!" Quarko cannot resist this quotable quip.

"Humor aside," Dabar says in a serious vein. "You have given this hydrogen atom a get-up-and-go attitude, Photie."

Elohim is ready to add a name to this definition. "You now live in an ionized hydrogen atom devoid of its accompanying electron. This is your new identity. You are now equipped to help the rest of your friends in a very, very creative project in this nearby molten mass, planet Earth. At present, you are all connected very strongly to its gravitational field. Our creation, space-time, alone will determine when this new phase of creation will begin." Elohim catches the attention not only of Photie but of all the atom friends.

"Again, I cannot wait to begin this new part of our creative journey, Elohim." As always, Quarkie is energetic and trust-filled.

"I trust you deeply, Elohim, Sophia, Dabar, and Ruah." Nitro is growing in maturity, courage, and responsibility.

Sophia tries to summarize their creative journey up to the present. "All that you have experienced to this point, my dear friends, has been only a preparation for possibly a more important mission."

There is a sense of wonder that envelopes this friendly group of atoms. They remain quiet and reflective.

I gaze back at the proto-sun forming in the womb of this massive nebula. I know that nuclear fusion ignition is close at hand. About 93 million miles away, this molten mass, planet Earth, continues to be impacted by large and small meteorites, asteroids of various sizes, zooming comets loaded with elemental and molecular products, and infinitesimally small grains of intergalactic dust and gases. As they enter this fireball, they are vaporized. They become a part of this inferno.

What will become of this mysterious, violent, red, flaming planet Earth? How will my atom friends relate to this Earth? Is it possible that new and novel creativity will arise out of this molten chaos?

Chapter 21
An Astounding Journey

At the heart of the proto-sun, the temperature at its core surpasses the magic 18 million degree mark. The nuclear fusion furnace ignites. This action releases a massive surge of photons that carry light and spectral radiation, which mushrooms through the masses of orange hydrogen gases that surround the helium 4 nuclei core. These photonic energy packages illuminate the sun's surface, coloring it yellow-white. Giant surges of flaming gases leap out of this fiery sphere in many places on its surface, agitating all of the newly forming planets.

It is very late on September 11, still about 10.4 byich, or to put it another way, 4.6 billion years ago (bya).

Photonic radiation from this newborn sun fills the billions of miles of intrastellar space in all directions. When this vast wave of energy meets the molten mass, the atmosphere of the Earth, now about 93 million miles from this new sun, is heated significantly.

In the atmosphere that surrounds this liquefied ball, the radiant energy assists in the formation of new molecules of carbon- and hydrogen-based compounds. Gravity binds these new molecules to the atmosphere surrounding this reddish-brown molten planet.

Instantaneously, my vision allows me to travel toward the sun. The intrastellar space of this newly born sun is littered with debris. I see two other smaller planets closer to the sun being formed out of this space-time refuse consisting of meteorites, asteroids, comets, dust, and gases. Some of these bodies of mass are several hundred miles in diameter; others scraps are much smaller. Gravity is gradually drawing all of this matter into two molten masses much like planet Earth. (I recall from my science class that these two neo-planets are called Venus, the larger one and closer to the Earth, and Mercury, smaller than both Venus and Earth, and closest to the Sun.) This absorption process is violent. Space junk continues to slam into these two new planets. The energy generated by the collision further liquefies and heats this fluid mass. The process is similar to a potter throwing new clay onto a ball of clay seated on a spinning wheel. The ball begins to grow larger and larger. But for now, these balls are liquefied and fiery.

I turn my attention beyond planet Earth. As time goes on, the intrastellar space-time is still filled with asteroids and planetesimals, but of much smaller size. Again, I see another planet in formation beyond the earth. The process of accretion continues similar to Earth, Venus, and Mercury. (I recall from my earthly knowledge that this planet is called Mars.)

My vision quickly brings me to dark, colder intrastellar space-time beyond Mars. I move through an area populated with asteroids, large and small, comets, and meteorites into a realm of the lighter gases. Clouds of gases, hydrogen and helium, are energized by the radiation. Being at great distances from the new sun, molecules of water, methane, and ammonia are frozen. Gravity binds these gases and molecules to form massive, gaseous planets.

Even as I travel effortlessly beyond the speed of light past the first gaseous giant, the latter is attracting and mopping up the gases and molecules to grow larger. (I remember this planet was called Jupiter.) Although this is a relatively cold, gas-filled planet-in-formation, the core is made up of helium surrounded by hydrogen compressed into liquid form generating heat. On the surface, violent winds blowing hundreds of miles an hour ravage the planet and paint its clouds in rich shades of yellow, orange, red, and brown. I notice that there are small satellite planetesimals circling in orbit around planet Jupiter. They also are mopping up small clumps of mass along with dust particles, which add to their size. The heavier elements sink toward the core of these small planetesimals. They are presently moltenlike at their core and gaseous at their surfaces. Jupiter's surface clouds intermingle fluffs of pink, rust, yellow, tan, orange, and white tones.

I traverse farther out into intrastellar space-time away from this newborn sun. About 900 million miles from the neophyte sun, I come upon another gaseous planet. (Again, from my earthly life, I recall that this planet is called Saturn.) Gravity is attracting the light gases, hydrogen and helium, along with a smattering of heavier and more complex molecules. Saturn is subdued and painted in drab, yellowish brown tones. This planet has the beginning of a series of thousands of rings orbiting its center. Each ring consists of dirty ice ranging in size from tiny grains to large house-sized clumps of ice. As with Jupiter, Saturn has the beginnings of small planetesimals circling it. Most of these satellites are comprised of water, ice, and some rock.

My journey continues outward toward another neo-gaseous planet. (I remember its name, Uranus.) Uranus is about 1.8 billion miles from its newborn star. Again, it consists mainly of hydrogen and helium gases. Traces of methane gas inhabit its thick cloud cover, which absorbs the red, yellow, and orange colors. Uranus projects back to its viewer a blue color. About a dozen very tiny ringlets, consisting of ice, rock, and dust, circle its equator. Uranus is unusual as its axis is almost horizontal to the plane of its orbital motion around the sun. Again, I can see the beginnings of small planetesimals being bound by gravity into varying orbits around this planet.

Here I am, 3.1 billion miles in the cold, bleak intra-planetary space-time, approaching still one more gaseous planet. From a distance, it seems placid and inactive. As I approach this gassy sphere, it is covered with clouds, but not as thick as on Uranus. A jet stream drives its clouds thousands of miles an hour from east to west. As seen from a distance, the methane gas in the upper clouds colors the planet slate blue. (Again, I recall this eighth planet is called Neptune.) A very active ice planetesimal orbits Neptune, along with several other nascent bodies of mass.

On the outer reaches of this amazing sun system, I discover the smallest planet. Temperatures of minus 450 degrees shiver the surface of this ice planet. A thin methane atmosphere gives a yellow hue to this rocky ice sphere. Bound by gravity is one planetesimal. (From my previous study, I remember that this planet is called Pluto.) The latter orbits about 3.8 billion miles distant from this system's core, the recently ignited yellow-white star.

My vision's ability to span huge distances in an instance allows me to stand at the edges of this solar system. From a vantage point on its far outer edge and slightly above, this system appears to be situated on a vast plain. It is like a gigantic phonograph record. Etched on its surface are nine elliptical rings, which outline the varying orbits of each of these planets around its neophyte sun. Only two orbits intersect each other. At one point, the smallest and outermost planet, Pluto, moves inside of Neptune, which assumes for a time the distinction of being the planet farthest from the huge, yellow-white sun at the center of the planetary system.

Within the Milky Way galaxy, this entire sun system is situated along one of the spiral arms surrounded by multitudes of other stars. This arm, along with the other arms and all of their hundreds of billions of stars, is rotating around the core of this galaxy. At the same time, the entire Milky Way galaxy is moving and expanding into space-time in relation to all the other billions of galaxies in the universe. Gravity is the power that is keeping every part of the universe united in a web of creation.

I am astounded as I pause to reflect on my journey from September 1 through September 11, 10.0 through 10.4 byich or

5.0 to 4.6 bya. I began by entering into the vicinity of the Milky Way galaxy, passing by its cluster of outer galaxies. I met my atom friends and the Divine Beings, surreptitiously joining them in their tour of this spiral galaxy.

What amazes me are the distances in both intrastellar and intergalactic space-time. The distance of 93 million miles from Earth to our neo-star is gigantic. Even the distance of 3.8 billion miles, from the Sun to the rim of this star system, is only a tiny fraction of a light year. Aside from our Sun, the closest sun to planet Earth is about five light years away. From Earth, this star appears as a tiny light in the distant intrastellar space-time. In one of the Milky Way's spiral arms are hundreds of millions of stars, of which our newborn sun is just one of the smaller ones. Yet these suns are just little specks of light when viewed from Earth by my eyes. This sun system is located forty thousand light years from the core of the Milky Way galaxy. A colossal distance! Describing and imaging these fantastic distances is impossible. How is it possible to describe the astounding distances between the billions of galaxies in our universe? I am rendered impotent in describing such an unbelievable, magnificent universe.

As my vision transports me back to the vicinity of this accreting molten mass, planet Earth, I gaze at this very tiny body and wonder, can this possibly be the site of tremendous creativity, as the Divine Beings claim? Why not this newly born sun, or one of the larger, cold, gaseous planets as a place of special creativity? They are larger and seem more significant than planet Earth. And this motley band of atom friends and all the other atom families, can they possibly be the source of new astounding creativity that the Divine Beings indicate?

For the first time, I begin to have doubts about the future. I want to live in hope. But molten, fiery planet Earth as the source of amazing creativity? Preposterous!

Chapter 22
Catastrophe Averted

The nascent yellow-white star at the center of this star system is now casting a bright glow upon this red, orange, and yellow liquefied planet Earth.

Chaos continues to reign. Gravity is allowing the earth to attract slate-gray asteroids and meteorites of varying sizes, gases, and dust to meld with it. All of this debris is crashing into the flame-red molten planet. It continues to grow in size.

The planet is beginning to create an atmosphere of dark, musty clouds of carbon dioxide, methane, ammonia, and water. These clouds are being fed initially by molecule-laden vapors generated by the fiery, liquid surface of the earth.

The heavy metal deposited by the many types of space debris sinks toward the core of the earth. This center, which is composed of members of the iron and nickel atom families, is heated to thousands of degrees. A traumatic event is ensuing. I notice that a planetesimal about one-quarter of the size of planet Earth is on a collision course with the latter. If these two large masses impact head-on, it will be a mammoth catastrophe for both. They threaten each other's existence as planet and planetesimal.

As the smaller body approaches, I notice that its course has changed. It now appears that it will only strike on the outer molten mantle of planet Earth. This planetesimal grazes the rim of Earth's molten mass, sweeping up some of its reddened plasma, which adheres to its surface. Then it rebounds away from Earth. Will it be freed from the gravitational tug of Earth and move out into intrastellar space-time? It is about fifty thousand miles, then one hundred thousand miles distant from Earth. But it seems to be slowing down as it continues its flight. By 175,000 miles, this mass has slowed considerably. Earth's gravity gradually recaptures this planetesimal, which settles into an orbit about two hundred thousand miles distant.

* * *

The time is September 13, still 10.5 byich or 4.5 bya.

Just above the dark, circling clouds, I see the Divine Beings, radiant in brightness, exuding their deep energy of love. The ever-present Ruah hovers lovingly over the other Divine Beings. As I approach closer, the atom friends come into view. They are immersed in conversation.

"I am transformed!" Quarkie is filled with enthusiasm. "Our carbon atom was caught in the waves of photonic radiation caused by the ignition of the fusion furnace in the new sun. We were swept into the path of a hydrogen atom. We found an immediate kinship with him. But that was not the end of our odyssey. Energized by this radiation, we were attracted to a nitrogen atom. Together, I feel we are a new identity."

"One atom of carbon, one of hydrogen and one of nitrogen create a special chemical bond and compound, Quarkie." Elohim begins to explain what has transpired in the wedding of these three atoms. "I give your new family a name. I name your new compound a hydrogen cyanide molecule."

"Hydrogen cyanide!" The "quick-change artist" Quarko seems concerned for once. "That name sounds sinister. When I think of it, the words *poisonous* and *pungent* come to mind."

"Does that not sum up your personality, Quarko?" Dabar begins to spar with him, gently taunting him.

Quarko abandons his serious mood and assumes a more mild tone. "So you are defining me as a queasy, qualmish quark." He lightens up and provides his own self-definition. "Really, I am a quintessential quark, in his purest form."

"Then you will add great dignity to that sinister compound, hydrogen cyanide, Quarko." Dabar admires his quick wit. "And you may also be prophetic, because in your new identity, you, Quarkie and Quarkoff, may be able to be creators of astounding new beings altogether different from that of the atom families."

With glowing pride, Quarko responds, "That lofty creative mission and destiny is why I am called the quintessential quark."

"Don't forget us, Quarko." Quarkoff is upset with the latter's boastful attitude. "Quarkie and I, along with our new hydrogen and nitrogen atom friends, are just as important as you."

Hydrojean is quick to remind the quark family. "We are a web of creation, a community. We are to cooperate in our creative mission."

"Well spoken, Hydrojean." Sophia affirms her insight. "*Creator* is a term that means what it says. You work in harmony and unity to accomplish your mission."

Phosie poses a question to Elohim. "You need to give some names to two important beings in our Sun system. The first is the new blue-white sun. The second is the planetesimal that almost destroyed our planet. The one that struck our Earth a glancing blow but is now held by gravity in orbit around our Earth."

Elohim begins to give names to these two beings. "We have already named these blue-white and yellow-white bodies, suns. We will call your sun, the Sun, and the Sun and its nine planets, the solar system." Elohim hesitates. Then he continues thoughtfully. "We will call the glancing planetesimal a moon. I rather like the word moon. We will call this mass that circles planet Earth, the Moon. We will also refer to any planetesimals circling any of the other eight planets in the solar system, a moon."

"I like the names Sun and Moon, Elohim!" Phosie expresses her gratitude to him. "They are short and pithy names. Thank you."

"Where do we go from here?" Nitro is inquisitive and anxious to get on with the creative mission.

"Where do we go from here, Nitro?" Sophia intuits his anxiety and concern. "We must all wait in patience."

The atom family listens to Sophia intently. They resolutely turn their attention to the future.

* * *

The time is September 15, 10.6 byich, 4.4 bya.

Stray gray asteroids and meteorites, icy blue water-laden comets, dust, and gas continue to impact molten planet Earth, but the number of accretions is diminishing. The nine planets, along with their accompanying moons, are mopping up most of the debris present in the intrastellar area of the neophyte solar system. The Sun, having just begun its life as a sparkling star, is growing in intensity of light and radiation, bathing the nine proto planets.

The four fiery planets closest to the Sun are losing their red glow and taking on a deeper maroon color. The very heavy members of the atom families, like iron and nickel, are sinking into the core of the planets. These centers are being fired by the volatile radioactive elements captured from the debris plastered onto the planets by the crashing space rubbish.

So Ruah was truthful when she told the atom friends that their own atom families and all the other atom families must ready themselves for the next "leap forward." In addition, they may help to develop all the other eight planets orbiting within the platter plane of the solar system.

I zero in on planet Earth. All over its maroon, molten, spherical face, huge volcanoes are spewing liquid magma high into the atmosphere. These belching geysers are gradually paving its surface as they solidify. In addition, they are seeding the upper atmosphere with carbon dioxide, nitrogen, ammonia, formaldehyde, carbon monoxide, hydrogen, and methane. As yet, I do not observe noticeable traces of oxygen gas in these dense chemical clouds.

* * *

The time is now September 22, 10.9 byich, or 4.1 bya.

The Earth is beginning to cool. Debris smashes into the planet with much less frequency. Fiery volcanoes are still erupting all over the Earth's surface. Being cooler, the dark clouds, rich in various molecules and gases, some including the atom friends, begin to condense. These gases, including fluid water, begin to fall upon its curved surface, cooling it. A hard, rocky crust is beginning to form. But as soon as the elements coalesce and harden, an earthquake occurs, a volcano erupts, or spatial junk from intrastellar space-time crashes into this crust and disrupts its formation. As time goes on, there is the beginning of a thin, hard surface on about one-tenth of planet Earth. The other 90 percent is filling with the rain waters and molecules deluging the Earth from the upper atmosphere.

* * *

The time is now September 27, 11.1 byich or 3.9 bya.

The earth has cooled appreciably during these past five days. A hard crust is continuing to form on its rounded surface. The chemically saturated clouds have continued to condense, flood, and fill the massive fluid basins with a rich molecular soup of water and various chemical compounds. Violent volcanoes belch their fiery red mass into the atmosphere to continue to seed the upper air with more and varied chemical molecules. These flaming geysers also continue to pave the crust of the Earth with a rich, liquefied carpet that soon hardens. The incessant rains pour down, continuing to fill the fluid basins that have been created by the hardened magma from the volcanoes. The rich deposits of carbon- and hydrogen-based compounds are being deposited by the massive rainfall into the gradually filling liquid basins. Continuous thunder echoes and rumbles. Awesome lightning bolts crackle and crash over the rock crust and fluid waters of planet Earth.

At present, the seas are a chemical soup rich in many different colorful compounds and simple molecules. The ultraviolet radiation from the Sun intensifies. Violent rainstorms, accompanied by repeated bolts of lightning and monstrous thunder, rage for tens of millions of years. In this dynamic environment, new and more exotic chemical compounds come into existence. Especially in the waters near the edges of the basins and ponds, the natural alchemy of the atom families is especially fertile and creative. Also, in the depths of the water basins, fissures in the rocks allow red hot magma from the massive volcanoes to enter the deep waters and pave the floor of these fluid basins.

The rains begin to wane. They have washed the atmosphere of many chemical substances and have rendered the air thinner. The waters now cover more than 90 percent of the surface of the earth. As yet, there is little or no oxygen in the air surrounding the earth. This planet is beginning to take on some color: browns and grays on the rocky surface and brackish yellows, reds, and oranges in the chemically laden fluid basins.

* * *

The time is late October 1, 11.3 byich, 3.7 bya.

Over the past four days, there are few visual changes on the earth. The crust of the earth has hardened and deepened. Volcanoes, earthquakes, electrical storms, and lightning bolts have lessened in frequency. Craters are now evident on the surface of the earth due to the violent crash landings of occasional asteroids, meteorites, and comets. The crust of the planet is painted in rich tones of browns, grays, whites, and pinkish reds. The water basins are brackish and pungent, still colored in dull tones of yellow, orange, and red. It is evident that on the edges of these basins and in nearby warm ponds of water, carbon- and hydrogen-based compounds have become more sophisticated and complex in character.

The atom family members, whose story I have been telling, have been carried by the rains from the upper clouds to the edges of the water basins. They seem to have acclimated themselves to their new home, the earth.

I pause my observations to muse on this more tranquil scene. This is probably the locale for the next burst of creativity, I think to myself. Ruah spoke on September 11 about some "great leap forward" in creativity. She further added that there was a need to prepare for this momentous event.

The earth home of the atom family members has gone through a great development, from a fiery, molten mass to a more placid planet with a solid crust and a very large fluid basin. This has been due to the ninety-two atom families who have worked to create the earth in this manner. Ruah has been very active in helping them to cocreate this planet Earth.

Are these preparations far enough along for the next leap? I again remind myself of the need to be patient and trust the creative process.

Ruah! Yes, she is nearby, brooding over this area of the Earth. I trust that this is the time. My wait is over. I await the future with exuberant expectation!

Chapter 23
The Atom Friends Land on Planet Earth

"I have changed! I have a new identity! I can feel the difference." Nitro is excited, yet befuddled. "But I am scared. I am attached to three members of the hydrogen family."

It is October 5, about 3.6 bya.

All of the atom friends, Quarkie, Quarkoff, Quarko, Hydrojean, and Nitro, have been carried on the wings of the powerful winds and torrential rains sweeping planet Earth. They have landed on a watery shore, which is now relatively cooled and hospitable for complex molecules to reside. The Divine Beings are present to them.

Sophia is nearby to console and affirm Nitro. "From the beginning, all the atom families have chosen to form communities of beings. You have been creators." Nitro is still in a state of disorientation. Sophia turns toward the quark trio and continues. "Do you remember that you, Quarkie, Quarko, and Quarkoff, have changed your identities several times, but you have held onto to the personal identities you possessed from the time of the creation of the quark family?"

Quarko smart-mouths Nitro. "So you are trying to steal my thunder as a quick-change artist." He smirks. "You are frightened! Where's all that bravado about trusting the Divine Beings and trusting in the process? Getting cold feet, nervous Nitro?" Nitro's fear is condensing into anger.

"Cool it, Quarko." Dabar is stern. "This is Nitro's first change of identity. You were frightened when you first joined the Helium 4 nuclei family. Later, under the intense pressure of the first imploding supernova, you melded with two other helium 4 nuclei to fashion the first carbon atom. Now, you are part of the odious (I use your definition, Quarko!) hydrogen cyanide family of molecules. So do not be so critical and sarcastic."

"That's right, Dabar." Nitro's anger subsides due to Dabar's affirming words. "It happened so suddenly. I was taken by surprise. When the three hydrogen atoms surrounded me and attached themselves firmly to my bonding sites, I was indeed frightened. I thought to myself, will this be my demise, like Hydro? I am reassured by your calming words." Nitro takes the opportunity to counterattack Quarko. "So you were very frightened at the time of your first change of identity! Courageous Quarko, fearful?"

Elohim has heard enough of this banter. "Let us stop this now!" He is very emphatic. "The most important part of your new identity, Nitro, is that you have a new name. You now are part of a gaseous substance called ammonia. Three members of the hydrogen family have joined you to create this new substance. You are so constituted, Nitro, that you have special sites that allow you to link up with these hydrogen atoms. Quarkie and his two friends, part of a carbon atom, also have special sites that help them bond with hydrogen, oxygen, or nitrogen atoms. With the quark trio, you are taking a further step in bringing forth the big leap forward, which Ruah has prophesied."

Elohim's words have no sooner been concluded when Hydrojean comes on the scene. Excitedly, she proclaims, "I have joined a new relationship with two hydrogen atoms and an oxygen atom. Isn't that wonderful?" Nitro and the quark trio are speechless. In the past, she would be frightened by a new identity, but Hydrojean is enthusiastic. "I need a name for my new relationship."

Sophia empathizes with her. "You have picked up the creative spirit, Hydrojean. You are pushing the creative process forward with enthusiasm." Sophia ponders momentarily before she responds to her request. "We will call your new being a molecule of formaldehyde, a gaseous creature."

"I like this new name, Sophia. It's a formidable name." Hydrojean then enunciates her new name. "Formidable formaldehyde! Now that's a name! I am proud to be a part of this new gaseous substance."

Quarko refuses to let this opportunity pass. "So, confident Hydrojean is proud of the name for her putrid gas. Our hydrogen cyanide is at least a pungent gas. There is a difference." He ends with a note of superiority.

Dabar reprimands Quarko. "You are always stirring up controversy." He steers the conversation to another subject. "Do you notice, Quarkie, Quarko, Quarkoff, Nitro, and Hydrojean, that there are five atom families that are involved in the creation of these new substances?" He pauses, then adds, "Let's call these substances, molecules."

"No, I have not noticed with whom we are allied in these new molecules," Hydrojean responds with rapidity.

"The hydrogen, oxygen, carbon, nitrogen, and phosphorous families are the five of which I speak." Dabar places emphasis on each family. "There is a sixth one that will be important later on in your creative work. This last one is the sulfur family."

"You mentioned the phosphorous family." Nitro is inquisitive. "Where is our friend, Phosie?"

"She will be here soon and be involved in the next phase creation," Dabar reassures him.

"I'm excited to move on into this next phase." Hydrojean is effervescent.

Quarkie has been silent since Quarko's early diatribes. She seems pensive. "The way you present it, Dabar, this will be the most momentous event in our 11.4 billion years into creation history (byich). I'm not sure that I can survive yet another overwhelming change in my existence."

Sophia has been observing Quarkie's demeanor. She addresses her. "Quarkie, we are probably asking a great deal of you again. We have tried to prepare you for each new traumatic, yet creative event, even though we do not know exactly what it will be. We just believe that you will continue to push the creative process to birth new, inventive beings."

Quarkoff has been silent. He identifies with Quarkie. "At times, I have found change terrifying. Especially in the core of the first supernova. When we, three helium 4 nuclei, came together to become the first carbon atom, I felt lost. Particularly when the white-hot core of this red giant exploded and we were impelled at blinding speed into the Milky Way's intragalactic space. I thought it would be my demise. It was not until our speed lessened and we were free of the chaos of that supernova that I could relax and be reoriented to my new existence. So I understand Quarkie's present anxiety."

"Thank you for sharing your experience, Quarkoff." Sophia pauses briefly and then readies herself to impart some words of wisdom. "Dear creator-atoms, as I mentioned, I intuit that you are about to enter a crucial creative phase. Ruah will be even more intimately present to this new creative process than she has ever been in the past. Even though her presence will be intimate, we still depend on all of you to assist our universe, and especially planet Earth, to achieve new heights of creativity."

Quarkie has been attentive. "So you are asking us to put our fears aside and move creatively into the future with Ruah, our directress." Quarkie then concludes, "What are we to create, oh Divine Beings?"

With honesty, Elohim addresses her question. "We do not know. We can only affirm that we had no definite plan in the beginning. We sent Ruah to breathe forth our energy of love to give birth to space-time. Out of our act of ecstatic love, you came into existence. From the outset, we wanted you to create. We wanted to fill this new universe with a community of beings with whom we could relate and share our unlimited love." The atom mates listen attentively. After a brief pause, Elohim concludes. "So we do not know exactly where our mutual cooperation will take us, as Sophia has already told you. But Ruah feels strongly that with all of you changing basic identities and with the five atom families now working so closely together, something special is about to happen."

"So you also do not know exactly what will transpire, Elohim." Quarko is serious for once. "All the recent changes in our identities and the change of the planet Earth's environment seem to indicate that a momentous creative event is about to happen."

Ruah has been a silent participant. She speaks. "Yes, you have noticed that fewer and fewer ebony asteroids, smaller golden tan meteorites, and icy blue comets are colliding with planet Earth. Black, mountainous volcanoes are spewing their reddish white hot magma onto to the ground to increase its black surface, while underwater volcanoes belch forth their crimson hot magma, which quickly cools and paves the floors of the expanding basins of water. Notice the atmosphere over the earth at the edges of the watery basins where you lie. They are rich with substances: hydrogen cyanide, methane, ammonia, formaldehyde, carbon dioxide, and water. Do you notice that the carbon, oxygen, hydrogen, nitrogen, and phosphorus families are intermingling more and more and setting up closer, more intimate relationships? This continues as I speak. There are still violent electrical storms full of energy and continuous torrents of water condensed from the cooling atmosphere filling the growing paved basins of water all over the earth. This environment seems ready for the next step in creative history."

Elohim, Sophia, and Ruah's honest appraisal causes the atom and quark characters to take heart and renew their trust in the Divine Beings and their unplanned adventure.

Nitro, whose cry of fear began this conversation, pledges his loyalty to the Divine Beings. "I believe that I can say for all of us that we will trust and remain close to Ruah."

* * *

I have been engrossed in this dialogue between the Divine Beings and the atom characters. I too have been observing the atmosphere and conditions on the surface of the earth. From September 27 to October 5, or about 300 million years in time, the interactions between the five atom families have quickened. Larger and larger amalgamations of these families are producing more complex compounds. This is happening near the edges of the waters and the nearby tidal ponds. The energy from the Sun and gigantic bolts of lightning that reach from the top of the sky to the edges of the water basins and ponds energize this breeding ground of new complex compounds.

I can see the same furious activity happening to the five atom families near the deep ocean vents, which spew forth hot steam, fiery red magma, and saffron colored sulfuric acid. These conditions are ripe for creativity.

These are the imminent signs of the "great leap forward" Ruah prophesied to the atom families on September 10, 4.6 billion years ago (bya) at our Earth's beginning.

With a heightened intensity of love energy, Ruah is brooding over the Earth.

Chapter 24
A New Creation Is Birthed

The translucent energy of love emanating from Ruah touches the slimy shoreline of these brackish watery basins and nearby pools. A lilting introductory adagio, reminiscent of Brahms and hinting of what is to come, accompanies this vivacious love energy. This most joyous, exuberant melody exudes from the eighteen vibrating strings and loops of strings living within each of the quarks making up the surrounding atomic protons and neutrons. This amorous melody vibrates and moves relentlessly into a soaring hymn of joy. This transpiring event is filled with a spirit of awe and expectation.

The atmosphere is dense with hazy gray carbon dioxide. The invisible ultraviolet rays cook the ocean waters and warm the surrounding air. Torrential rains continue to deluge the earth with water while tumultuous thunder and electrifying bolts of lightning pierce the waters and skies. At a level beyond the purely physical, Ruah's radiant touch vivifies and energizes the already long chains of carbon and hydrogen based compounds residing near the edges of the water basins. The accompanying amorous melody reaches a majestic crescendo in this energetic environment as the five atom families—carbon, hydrogen, oxygen, nitrogen, and phosphorus—are busy creating more and more complex substances.

At a profound level, Ruah's musical love energy seeks out the black crevices and blowholes on the floor of the water basins where hot steam, magma, and hydrogen sulfide shoot up from the fiery depth below. The conditions are perfect for Ruah's vivification of the long chains of complex sulfur-, hydrogen-, phosphorus-, oxygen-, nitrogen-, and carbon-based compounds that reside there.

In both of these locations, on the physical level, the six atom families, and on a deeper level, Ruah, are creating beings astoundingly different from the atom families themselves or the complex compounds already created within this gigantic universe and on planet Earth. In particular, a new creative process is being perfected. Will a new creation emerge from this effort? Only time will tell! In the background, a magnificent hymn of joy soars to greater heights of melodious musicality!

The time is October 5, 3.6 bya.

I must inform you that, as a visionary, my gifted vision has mightily magnified these microscopic locations. This allows me to observe the new micro-beings, the long chains of molecules, in such a way that I can see clearly the individual atoms and the elongated connections that hold these substances together. I am astonished by the creativity of the five atom families, spurred on by Ruah's radiant love.

Near this particular molecule-infested seashore, the Divine Beings are immersed in conversation with Quarkie, Quarko, Quarkoff, Hydrojean, Nitro, and Phosie. The latter atom has recently joined them.

Quarkoff initiates a dialogue. "Since October 1, we have been very, very busy. The cooperation among the five atom families has been amazing. In this very hot water, amid loud ponderous noise and powerful bolts of radiant light, our hydrogen cyanide compound joined other members of our five atom families in long lines. We tried new types of bonding. Being a carbon atom, we were in the center of activity. We could open binding sites to five atoms. I could see the same activity going on all around us."

"It was almost like we were in a competitive contest to see which group of atoms could construct the longest chains of compatible atoms," Quarkie opines. "Some of our efforts didn't seem to work out, so we would break up our chain of atoms and try another grouping. We seemed to intuit whether the formulations would work or not."

Dabar comments, "What did you other atoms experience?"

Nitro shares his experiences. "Our ammonia gas joined forces with other compounds made up of the five atom families. We also experimented with combinations extending our bonds to more and more atoms. It was exhausting work. But the community of the five atom families was enthusiastic in their efforts. I was energized too. I wasn't frightened."

"Nervous Nitro is gaining confidence. Wow! I congratulate you, my friend." As usual, Quarko spouts a biting rejoinder to Nitro's enthusiastic tale. "Seriously, I was energized too by this creative activity." Quarko is earnest for a change.

Hydrojean has not lost her enthusiasm. "When you urged us not to be fearful and to enter into deeper relationships with the other atom families, I did not realize that it would involve such an expending of cooperative effort. But it was worth it. My experience being a part of formidable formaldehyde was very close to that of Nitro. Much experimenting, some failures, but many successes. We developed many new substances."

Sophia spots Phosie. She was absent during their dialogue on October 1. "Good to see you again, Phosie. What has happened to you since we last saw you?"

"My existence has been much calmer most of the time since our last encounter." Phosie seems subdued as she begins her story. "I have seen the vast changes and developments on planet Earth. I have felt that Ruah's prophecy of a 'great leap forward' was close."

Phosie's demeanor changes abruptly. "My lethargy departed the moment that Ruah's love energy touched me on the edge of this watery shore. In this liquid environment, I was energized. I began to be more involved in the furious activities going on in the atom communities. I began to make some connections." Then she pauses momentarily. Her emerging buoyant manner causes everyone to center their attention on her. "But these connections led to an astounding insight and surprise. I became a part of a liquid bubble. Presently, all of you live inside of our sphere, doing the work you describe. Many members of my phosphorus family are involved in this bubble-making. Our bubbles are the outer walls within which all of you live in a semi-fluid environment. Our walls have openings so that other groups of atoms and molecules are able to enter and leave. This is happening within some mysterious type of yellow-brown material."

Elohim explains to Phosie this unusual substance. "Although the yellow-brown material is a mysterious medium for your creativity, it was the perfect condition needed." Elohim ponders for a moment. "I should give this material a name. I will call it clay."

"Well, the clay medium is a very fine home for our creativity." Phosie looks at the Divine Beings and her atom friends and continues. "Within these layers of clay and inside our walls, we have given birth to a new creature. It is much different from any beings with which you or I have ever been associated during these past 11.4 byich. Its birth was accompanied by joyous music of rich tonal quality, which often in the past has accompanied the presence of Ruah's love energy."

Quarko is skeptical of Phosie's conclusion about the existence of a new being quite different from all the creatures that have existed up to the present. "Are you sure that our bubble and its inhabitants, members of our five atom families, constitute an altogether new being? Music accompanying its birth? I believe you are sorely mistaken in your conclusion." Quarko is emphatic in denying Phosie's claims.

"This is a new being, Quarko! I am not mistaken." Phosie is strong in her own defense. "The only question I have is: what is my relationship to this new creature?"

Elohim has been listening intently. He hears Quarko's rejoinder to Phosie's observation. He is ready to answer Phosie's query. "Phosie, you are right in your observation and conclusion about the birth of this new being and the magnificent music that accompanied the presence of Ruah's love energy. You have co-created a new creature. Not only you, Phosie, and the other atom family friends within your bubble wall, but also you, Quarko, Quarkie, Quarkoff, Nitro, and Hydrojean, have also cocreated this new creature. All over the planet Earth, at the water's edges and in the depth of these watery basins near the volcanic vents, a whole new class of beings has been created by you and Ruah. At a very deep level, her process of energetic vivification has taken place. The radiant energy of Sun, the chemical reactions in the torrid sea water, the powerful lightning storms, violent heat near the volcanic vents at the bottom of the sea, and your persistent, creative efforts within this new being have all played a part in birthing this new class of creatures." Elohim's gaze creates an intimate bond with his listeners. Then he says, "The final source of energy for this birthing was the power of Ruah's translucent love. This love penetrated you, all the members of the five atom families, deeply within your beings. Again, your efforts were crucial and very important in cocreating this new group of beings."

"What name do you give to this new being?" Phosie presses Elohim.

Elohim deliberates momentarily and announces. "The name of this new being is Prokaryote Bacterium. For short, let me call this new creature Prokye 1. Its physical makeup will be called a body."

Elohim again pauses. "You were correct in your intuition, Phosie. This being differs from you. It possesses what I call biological life.""

"How does Prokye 1 differ in nature from us?" Quarkoff inquires.

"If you do not mind, Elohim, let me answer Quarkoff," Dabar interjects.

"Go right ahead, Dabar." Elohim yields to Dabar.

"When Ruah's radiant love touched the bubbles, each containing the several hundred of long, chained compounds, she gave life to Prokye 1 and all the other prokaryote bacteria that have been created in various places on planet Earth." Dabar allows the atom characters time to assimilate this new knowledge. He goes on. "Do you remember, my quark friends, when the temperature reached 18 million degrees in the core of the first sun? What happened?"

Not surprisingly, Quarko is the quizzical quark to respond. "All of a sudden, the core ignited in a radiant glow that mushroomed through the entire gigantic sun to radiate its light and energy into the universe."

"Something like that happened to Prokye 1." Dabar speaks with deliberation. "Prokye 1 was turned on. Because you live inside of it and have provided it with all of these wonderful compounds, it is now able to generate its own energy. It does not need the warm waters of the basins, the lightning bolts, or Ruah's touch to keep it going. Through your experimentation on the physical level over the last few days, through your failures and successes, you have developed the means whereby Prokye 1 can live on its own," Dabar concludes. "Can you now understand, my dear atom friends, how you and the five atom families are cocreators with Ruah, Elohim, Sophia, and me? I am deeply proud of you."

139

Quarkie is confused. "But how do we relate to Prokye 1? We live in it. Does that mean that it controls us? Can it dominate us? If it can, I would be very, very frightened about the future of our relationship." Her feeling of anxiety permeates her words.

Sophia responds to her anxiety. "Dear Quarkie, have no fear. Do you remember when we spoke about how you were a community of beings bound together by the power of gravity Ruah established at the beginning of time?" Sophia measures her words carefully. "Prokye 1, our new being, is now a part of that community. It depends on you for its existence. If you do not manufacture its energy supply to keep it going, Prokye 1 ceases to exist, like our old friend Hydro. You will go on existing even if Prokye 1 is annihilated sometime in the future. If you work together, you both continue to exist. You are comrades. You need each other. So don't be frightened."

Elohim inserts himself into the conversation once again. "Quarko, you also asked another question earlier. You questioned Phosie's statement that she heard music accompanying this birth of Prokye 1. 'Music accompanying its birth?' you asked. We have never informed you that each proton and neutron is made up of eighteen vibrating strings and loops of strings. Your own strings and loops of strings were involved in creating the music that accompanied this new birth of Prokye 1. You need to become aware of your own musical nature, Quarko!" Elohim concluded.

"Wow! I am a whimsical, melodious, quintessential quark!" Quarko exclaims.

"That is enough of your braggadocio, Quarko." Quarkie is tired of his boisterousness. There is a pause in the conversation.

Nitro breaks the silence. He wonders whether they can now rest from their creative work. "Is this the end of our creative work, Sophia?"

"This is not the end, but the beginning of an exciting new phase. Your creative skills will be tested again." Sophia issues a challenge. "Will you and Prokye 1 be able to cooperate so that you can help it to make mirror images of itself, a new prokaryote bacterium? That is the challenge I issue to all of you! If you fail to replicate Prokye 1, creativity on Earth will cease." Sophia's gaze pierces the atom characters with profound intensity. Then she says with confidence, "I believe deeply you will answer it."

Chapter 25
Prokye 1 Is Introduced to His Cocreators

"I want to meet this new being, Prokye 1." Quarko wants desperately to meet this new being he has cocreated. "Sophia, you told us that Prokye 1 depends on us for its existence." Quarko pauses and then continues. "That makes me more intrigued with Prokye 1. I definitely want to meet him."

The time is still October 5, 3.6 bya.

Sophia hears Quarko's inquisitiveness, but wonders if the quipster is serious. She responds quickly. "Quarko, and my other atom friends, let me introduce you to Prokye 1, the new creature that possesses biological life that you have cocreated with Ruah."

Quarkie greets Prokye 1 with enthusiasm. "Welcome, Prokye 1, to our universe and our planet Earth."

Prokye 1 is overwhelmed with confusion and questions at this greeting. "Who ... who ... are ... are ... you?" it stammers.

Sophia empathizes with Prokye 1's dilemma. "Rightfully so, you are confused, Prokye 1." She carefully selects her words. "I am one of the Divine Beings, Sophia. Let me introduce to you my Divine Companions: Elohim, my husband, Dabar, our son, and Ruah, the personal love that binds us as one. Over 11 billion years, we decided to share our overwhelming love and existence with new beings. We sent Ruah on this mission. She breathed forth our overwhelming love and created time and space. The new being, time, includes events of the past, present, and future. We called the space, the universe. Quarkie, Quarko, and Quarkoff (you haven't been introduced to the latter yet) were the first beings created. We called them quarks." Sophia pauses for a few moments to let Prokye 1 absorb this story. "The universe expanded stupendously in the first second. Gradually, it cooled down so that the quark trio and their families of quarks could become part of the hydrogen and helium atom families. In fact, we asked these new creatures to be creators in the universe. They and all of the other ninety-two atom families agreed. They have been involved for 11 billion years in populating this immense universe with vast communities of suns called galaxies."

Quarkie takes up the story, first introducing herself. "I am Quarkie, the first quark." She proceeds. "Prokye 1, you now live in the Milky Way galaxy in a solar system on the planet Earth. Just a very short time ago, the members of the five atom families—hydrogen, oxygen, carbon, phosphorus, and nitrogen, of which the three of us are members—worked to bring you into existence. Ruah touched us with her powerful energy of love to energize our creative work and touch you, Prokye 1, with her energizing vivification in the depth of your being. You sprang into biological life." She observes that Prokye 1 is perplexed by all of this new information. She pauses. "I know how you feel. When I came into existence in the intensely hot, dense atmosphere of the first second of creation, I was befuddled, stuttering and stammering out my questions. Being sensitive to my perplexed state, Sophia was deeply empathetic. Dabar informed me that he wanted me to be a creator in the universe. Then he asked me what I wanted to do. Gradually, I came to understand that the Divine Beings really loved and cared about me. With their reassurance, I told them that I wanted to join with other quarks to form a community. Dabar and Sophia encouraged me to follow my desire. That is when I joined Quarko and Quarkoff to create the first proton."

Quarkoff takes over the historical recounting. "Prokye 1, welcome to planet Earth. I am Quarkoff, a member of the quark trio."

Fear seems to be falling away from Prokye 1. "I am happy to meet you, Quarkoff."

"During our 11–billion-year journey, we, along with other members of the hydrogen and helium families, have been engaged in creating," Quarkoff continues. "When the first blue-white sun in our universe ignited, we were present at its core, helping to create this enormous being. A little later, when this first star grew old, it had very few members of our hydrogen and helium families to fuel its nuclear furnace. It expanded and grew into a giant red star. Suddenly, its outer gaseous shell collapsed on all of us remaining in the core of our fuel-deprived furnace, crunching us together. The three of us, along with two other helium nuclei, were transformed into the first carbon atom. In fact, in this collapsed red star, our hydrogen and helium families created the ninety-two other atom families. You will meet some of them as you journey with us into future time."

"But why do you live inside of me?" Prokye 1 wonders

Quarkie is quick to respond. "Since the beginning of time and our coming into existence, we have worked as a community of beings to create. Just about a billion years ago, after the Divine Beings had taken us on an epic journey to discover the wonders of our vast universe, they told us that we were destined soon for a special mission. We happened to be present in the vicinity of an exploding supernova, which gave birth of our solar system, its sun, and nine planets. This explosion ultimately caused us to land on planet Earth, the third planet from our sun. Our continuing mission of creating was to happen here. We have joined five atom families to work on this new project under Ruah's direction. Amidst torrential rains and powerful lightning bolts near the scalding waters, we worked mightily to develop new and longer chains of atoms.

Suddenly, within a bubble wall, which the phosphorus family created, we developed the right combination of molecules to create you, Prokye 1. Then Ruah touched you with her intense love energy. You, Prokye 1, were vivified. Thus, we completed our mission. We have also developed a source of energy, which will allow you, Prokye 1, to live on your own, of course, with our continuing help. In fact, we live in you. We are a part of you" Quarkie ends her tale. There is a pause. Then, as an afterthought, she says, "Remember, there are multitudes of other atom family members who also help us provide for your internal energy. All over planet Earth, other communities of atoms have also created numberless prokaryote bacteria like you, Prokye 1."

Sophia wonders why Quarko has been relatively quiet to this point. Yet, she knows that sooner or later, he will take over the conversation.

"My dear Prokye 1, I am Quarko, the quintessential quark." He is in a boastful mood. "If you run into any problems, I can help you. I have a little experience. Let me say modestly, more than 11 billion years. That is a long, long time. I am a master of information to answer your questions."

"You exhibit a very, very modest attitude, my dear Quarko." Sarcasm colors Elohim's words. He then turns his attention to Prokye 1. "Seriously, Prokye 1, do not let Quarko fool you. He is an experienced quark but tends to exaggerate his prowess."

Quarko is not about to let Elohim's description of him go by without a comment. "My dear Elohim, I am shocked by your appraisal. Did you not observe that I was the intelligence behind the development of the powerful molecules that Ruah was able to use to vivify Prokye 1?" He pauses and smiles at Quarkoff. "Isn't that right, my friend?"

Quarkoff has heard enough of his bravado. "Quarko, you are a big liar. We worked together as a community of atoms to bring Prokye 1 to life." He addresses his next comment to Prokye 1. "Be aware that our friend Quarko is a quirky quark."

Quarko is laughing heartily. "Did you not realize that I wanted to tease you? Can you not take a joke? Although I lied to you, I am still a colossal quintessential quark."

Hydrojean has been attentive. "Prokye 1, I am Hydrojean, a member of the hydrogen family. I too live inside of you. I welcome you to our planet Earth. With your birthing into biological life, Prokye 1, we have not ended our mission of creating." She looks to Sophia and asks, "What did you mean when you told us that we would be called upon to help Prokye 1 to continue the creative process? In fact, you issued this as a challenge, which you believed we would answer."

Sophia, full of maternal love for her creatures, responds. "Hydrojean, my beloved, just as you worked and experimented to create Prokye 1 and a source of living energy to enliven Prokye 1, now you must engage yourself in the quest of creating atom compounds that will allow Prokye 1 to create a mirror copy of itself, a new prokaryote bacterium. Otherwise, Prokye 1 will be your final creation."

Nitro and Phosie have listened with intensity. They also live inside of Prokye 1. Nitro is the first to speak. "If you remember, Sophia, I said that all of us would take up the challenge. I again affirm our mutual intentions. Truly, you can count on us."

"Yes, you can count on us, Sophia," Phosie says forcefully. Then she addresses Prokye 1. "Nitro and I apologize for failing to introducing ourselves. We also live in you. I am Phosie, a member of the phosphorus family, and this is my friend, Nitro, a member of the nitrogen family. All of us have been working mightily to create you."

Prokye 1 is beginning to feel at home on planet Earth. It intuits that all its cocreators really love it. It accepts the introductions. "I am happy to meet you, Phosie and Nitro. I am most grateful for all of you giving me life and existence on planet Earth."

Nitro seems perturbed. He says, with intensity of feeling, "Let us not forget Ruah. Without her powerful love energy that sparked us to extra effort and her vivifying touch, you, Prokye 1, would not be alive."

Ruah is very grateful to Nitro. "This is very thoughtful of you to remember my part in Prokye 1's coming into existence." She gazes intently at her atom friends. "We are cocreators. We need each other. Since this was such a momentous event, I acted more intimately with all of you." She then directs her attention to Prokye 1. "You are a special creation. I prophesy that from you will come an immense progeny. But only with the powerful assistance of the atom families who live within you."

Prokye 1 is overwhelmed by Ruah's prophetic words. "I feel so humble to be a part of this growing community on planet Earth. I pledge my fidelity to you, Ruah, also to you, Elohim, Sophia, and Dabar. I am ready to take up Sophia's challenge."

"I believe in your sincerity, Prokye 1," Sophia responds. "But it will not be easy."

With dismay, Quarkie complains. "Why will it not be easy?"

"Not every creature will complete this long journey. Do you not remember Hydro? His pride led him to annihilation." Elohim responds to Quarkie's question. He measures his words. "Not all will survive, but all will be enlivened and energized as they contribute to the mission. Every being that enters into the creative process will find deep fulfillment in assisting the creative journey of planet Earth."

Quarko is the first to respond to this ambivalent prediction of Elohim. (I wonder what the quipster will say.) "We have believed in you, oh Divine Beings, and have trusted you up to this point." He pauses. (I am waiting for some humorous punch line.) "I have no need to doubt that what you have just prophesied will come true. I also believe that you, Prokye 1, will indeed find fulfillment in carrying out your creative role." (I am shocked by Quarko's candor.) He continues. "Let us move forward now to meet the new challenge."

Chapter 26
The Creation of Prokye's Image, a New Prokaryote Bacterium

"We need to organize our efforts to carry out the mission Sophia gave us." Nitro is taking the initiative. He speaks to members of the carbon, hydrogen, phosphorus, oxygen, and his own nitrogen families, many of them living in molecules and other elemental compounds, within the semi-fluid environment of the bubble membrane of Prokye 1. He urges them to join him in developing chains of molecules that might help Prokye 1 to make a mirror copy of itself. "We need to try different combinations of molecules and other compounds to contribute to the accomplishment of our mission."

The time is October 8, 3.4 bya.

Nitro notices that within Prokye 1's fluidic environment, Quarkie, Quarkoff, and Quarko, as part of a carbon atom, are working furiously with the hydrogen, nitrogen, phosphorus, and oxygen family members to assemble a new elemental chain to contribute to the mission. "How is your effort proceeding, Quarkie?" Nitro shouts.

Quarkie pauses from her work to respond. "With the hydrogen, oxygen, and our carbon families, we're still experimenting. We thought we had developed a compound of atoms that we could contribute to our joint mission." She seems frustrated. "We just feel that our endeavors to-date are futile. But we continue to experiment."

"We are engaged in the same process and have had little success so far." Nitro is upbeat. "But we really believe that we will be able soon to accomplish our mission. Yet, whatever compounds we develop must possess the capacity to have imprinted on themselves the basic body design of Prokye 1's essential makeup."

Hydrojean enters the conversation. "I am with you, Nitro. I believe that the idea that you just outlined is a brilliant insight. It can help us to create just the right kinds of compounds." She ends her response to him. "Right now, I am helping the quark trio."

"I am feeling ambivalent, Nitro and Hydrojean." Quarkie wants to defend herself. She thinks that they believe that she has given up. "I am deeply discouraged, and yet I believe that in the end, we will be successful. Allow me the opportunity from time to time to vent my frustration." She turns to Nitro. "Your recent insight is excellent. It should help us in the development of just the right compounds."

"Do not worry, Quarkie; we understand how you can become very frustrated." Hydrojean tries to console Quarkie. Nearby Nitro nods encouragement.

Quarko has been eavesdropping. He enters the dialogue with gusto. "With me at your side, Quarkie, there is no need to fret. With my experience and leadership, this project will be easy." Quarko gloats about his superiority. "Trust me, Quarkie. Our part in the mission is about to be accomplished under my quintessential leadership."

"You egotistical, puny particle," Quarkoff responds with bluntness. "Quarkie has been leading us. You have been doing little to further our efforts." He further chastises Quarko. "Cease the loudmouth boasting and help us with our efforts. Begin by contributing some excellent ideas, such as Nitro has done."

"Nitro may have his insights." Quarko defends himself. "You just do not appreciate my quirky brilliance. My wholehearted efforts you shall have, my dear friends." Then he adds, "You can trust this eminently quotable quark, Quarkoff."

"I hope you finally double your efforts, Quarko." Quarkoff is stern. "So get to work and help us."

Phosie has recently been released from her position as part of Prokye's exterior wall membrane. Her place was taken by another member of the phosphorus family. Now she is energetically working with the oxygen family to help create a duplicate copy of Prokye 1. "I have been leading members of the oxygen family to produce a compound to contribute to our joint mission to help Prokye 1. I think we have been successful. Since I have four binding sites, I have invited four oxygen atoms to occupy these connections." Phosie hesitates. She carefully chooses her words as she continues. "Since I have led the effort, I hope you don't mind if we call our creation, the phosphate group. As soon as you are ready, we are here to join our group of elements to your chains of ingredients."

Ruah has been observing their tremendous efforts. "What a remarkable group of atoms and molecules you are. I believe your strenuous efforts will be able to help Prokye 1 create a copy of itself." Ruah then addresses Quarkie. "This work is frustrating. Your dogged effort, though frustrating, will be rewarded with success, I feel sure." Then she directs her next comment to Nitro. "Your insight is brilliant. I believe that all of you are on the right track to accomplish your mission soon."

"Thank you, Ruah." Nitro accepts her affirmation. At the same time, he appears very excited. "I think we have found the right combination of atoms. After repeated trial and error, working cooperatively with the carbon, oxygen, and hydrogen families, we developed not only one combination, but five different but similar molecules. I believe we have achieved our part in bringing about this new creation." He hesitates momentarily. "Like Phosie, I ask your understanding when I tell you I would like to call our compounds the nitrate group, since I have led this effort."

"You have chosen a good name, Nitro." Ruah commends him. "You don't have to be apologetic." She pauses and then continues. "We can call your group nitrogenous bases, since you have developed five bases that are slightly different from each other. Each of these bases will be called a nucleotide. They open up the possibility of using various combinations of these bases, which might help you to implement Nitro's insight."

"I think we have finally achieved our goal." Quarkie is exuberant. "We just made some adjustments at our binding sites. This compound seems to be the right one." She goes on to say, "What shall we name it, Ruah?"

Ruah ponders momentarily. Then she addresses Quarkie. "Let us call your compound the sugar group."

"That is sweet-sounding, Ruah." Quarkie's frustration has given way to exultation. She basks in the glory of their achievement.

"Once I put my full efforts into the mission, we were successful," Quarko boasts.

"Your effort was important, Quarko, but we work as a community of beings to accomplish our cocreative task." Quarkoff is more gentle than usual in his response. He too is aglow with their joint success.

"Now you need to use your sugar group to build a bond between the phosphate group and the nitrogenous bases." Ruah is encouraging them to finish their mission.

"Our group will use one of our five carbon binding sites to link up with Phosie's phosphate group." Quarkie continues in her jubilant mood." Then she pauses for a moment. She seems to have been enlightened. "Then we will use one of our carbon sites to bond with the nitrogenous bases. So our sweetness bonds our three groups together." Quarkie expresses a note of satisfaction.

"We have not fully resolved our problem of being able to reproduce a copy of Prokye 1." Nitro cautions that more creativity is needed to accomplish their mission.

"What do you suggest, Nitro?" Quarkie inquires.

"We now need to create various combinations of nucleotide bases and imprint on them the body plan for Prokye 1," Nitro advises. "From its body plan, made up of these strings of encoded nucleotide bases, a replica of Prokye 1 can be constructed. This includes the exterior membrane wall, the power system, and all of its other body parts."

"Ruah, what name can you give to the raw materials we will use to reproduce a replica of Prokye 1, working from the body plan of our encoded bases?" Hydrojean inquires.

"You have already used these materials in the creation of Prokye 1." Ruah begins a response leading to the naming. "In fact, these molecules were formed in the atmosphere during the early days of the formation of the Earth and were carried to the edges of the water basins. They were made up of a carbon atom, which has these groups attached to its binding sites: a hydrogen atom, carbon-oxygen-hydrogen group, a nitrogen-hydrogen group, and a side group. The latter group is something like five nucleotide bases in that there are twenty different side groups. Again, this allows the formation of numberless combinations of these molecules." Ruah allows this information to be assimilated by the atom friends. She then names the molecule. "I name these molecules amino acids. Again, there are twenty distinct kinds of amino acids."

"Who cocreated these amino acids, Ruah?" Quarkoff asks.

"Your relatives in the carbon, oxygen, nitrogen, and hydrogen families," Ruah responds. "Interesting. You used their creations to cocreate Prokye 1. I am now giving you a name to their creative work. The amino acids are the molecules that help you carry out the construction plan written upon the encoded nucleotide bases. When you begin the construction of Prokye 1's replica, the encoded nucleotide bases tell you what combinations of amino acids to string together to build the various parts of your new prokaryote bacterium. This encoded body plan contains all the instructions needed for building this new biological life."

"If I understand you right," Nitro says, beginning to interpret Ruah's explanation, "we first need to build the combinations of the five nucleotide bases, upon which we will encode the proper combinations of amino acids required to build each part and function of Prokye 1's mirror image. After that, we need to transfer these encoded combinations of nucleotide bases containing its body plan to our building site within Prokye 1's aqueous interior. Then, working from the encoded plans, we collect and string together the combinations of amino acids that make up the various parts of this new being. Finally, we bring these parts together." He pauses. "Then, new life. Mission accomplished!"

"Before we begin construction, you need to do some naming, Ruah. What name do you give to these various strings of amino acids?" Hydrojean asks.

Ruah says solemnly, "Let us call them proteins, the building blocks of new life."

"You also need a proper name that encompasses all the nucleotide bases, Ruah," Quarko interjects.

"These new molecules will be called RNA," Ruah proclaims. (I recall from my readings in a recent biology text that this abbreviation, RNA, stands for ribonucleic acid.)

"Can we start to build Prokye 1's replica?" Phosie is anxious to move forward.

Within Prokye 1's rich, liquid interior, Quarkie, Quarkoff, Quarko, Hydrojean, and Phosie, along with all of the families of atoms present in the various molecules, begin to arrange the combinations of nucleotide bases. They encode on the proper bases the names of the many amino acids strings used to build all of the parts and functions in the proposed new prokaryote bacterium. They move this RNA long string of nucleotide bases to the interior of Prokye 1. With great effort, from the RNA body plan, this whole community of atoms are laboring to gather and string together the prescribed amino acids and thus build all the parts, walls, functions, and semi-fluid environment of this prokaryote offspring.

Suddenly, Nitro lets out a shrill cry. All turn toward him. He is being drawn into this new creation. Fear permeates his being as he cries. "I am being enlisted to become a part of this new prokaryote bacterium. I do not want to leave you, my friends. I am terribly fearful."

About the same time, Hydrojean screams. "I, too, am being drafted to serve as a member of this new creature. I do not want to leave you, Quarkie, Quarko, Quarkoff, and Phosie." She is extremely anxious and distraught. "I am deeply fearful. Ruah, save me!"

The exultation over accomplishment of their mission gives way to consternation. They are beginning to grieve their loss of close friendships.

The atom community begins to see the mirror image of Prokye 1 being created. In fact, this new bacterium comes to life. The mission given by Ruah has been completed. But Nitro and Hydrojean are departing. Prokye 2, the new bacterium, passes through Prokye 1's membrane wall to assume its independence and individuality.

The atmosphere within both Prokye 1 and Prokye 2 is saturated with feelings of grief, loss, fear, depression, and anxiety. The community of friends is shattered.

Amidst her sense of grief and loss, Quarkie is still able to realize that there is still one more step in their construction process. She advises, "We have not guaranteed that we will be able to easily produce yet another replica of Prokye 1."

"What do you mean?" Quarko questions her statement.

"We have accomplished our mission, Quarko." Quarkie is gentle with him. "Unless we save the RNA body plan we have just utilized to build Prokye 2 with each new replication, we will have to construct a whole new RNA design plan made up of the encoded combinations of nucleotide bases." She hesitates briefly and says, "I suggest that we take the present RNA design plan we have just used and place it in its special location within Prokye 1. Then, when we set about cocreating a new replica, we can start from the RNA body plan we have saved."

"What an ingenious suggestion." Quarkoff accepts her proposal. "Let us set about to transfer the RNA body plan to a special home inside of Prokye 1."

They find a suitable home for RNA. Temporarily, their sadness is forgotten.

Chapter 27
Quarko's Suggestion Leads to a New Creation

"Ruah, where have Nitro and Hydrojean gone?" Quarkie is excited and upset.

Ruah is still close at hand. She calmly responds. "Sophia challenged all of you to help Prokye 1 to reproduce a copy of itself to continue the creative process. Nitro and Hydrojean have departed from Prokye 1 in order to become a part of Prokye 1's mirror image, Prokye 2, so that when we set about cocreating a new replica activity. They are lost to you now, but you are still bound in unity with them through my love energy, gravity." She concludes her remarks.

The time is October 9, 3.4 bya.

"We need them here to help us." Quarkie is insistent. "I miss them very, very much. We have shared much over these 11.6 byich."

"I agree with you, Quarkie. We need them here." Quarkoff supports her. "So we are connected with Nitro and Hydrojean by the love force of gravity. But that is much, much different than being present to them, atom to atom."

"I understand that you are grieving deeply your loss." Ruah tries to console them. She also wants to redirect them to the mission they embarked upon a long, long time ago. "I reiterate: they needed to leave Prokye 1 so that they can continue to push the cocreative process forward."

"Will we see them again, Ruah?" Quarkoff is concerned.

"I cannot give you a definite answer, Quarkoff." Ruah tries to be gentle. "Only the future of space-time will answer your question."

"Good riddance to inferior atoms." As usual, Quarko has his own opinion. "Both of them were fearful atoms. The creative process needs strong, fearless leaders like me. I hope that's the last time we will be present to them. Yes, good riddance."

"You insensitive, egoistic, quirky quark!" Quarkoff is angry. "I'm just sorry that I am bound to you through the strong love force that binds quarks like us together. Unfortunately, in the first second of creative history, Quarkie chose you and me to be her companions through this long journey. I would like to get a divorce from you, Quarko."

"You are too community-oriented, Quarkoff." Quarko is not apologetic. "The creative process needs decisive leaders like me to cocreate."

Ruah intervenes in the conversation. "Let us put this argument aside. There is much more cocreativity ahead of all of you." Then she addresses Quarko. "You have tended to be a source of disunity all the way through creative history. Your humor at times has kept everyone loose during crises. At other times like this, you are creating barriers to creativity and unity. You will have to mend your ways."

"So our valiant energizer is upset with me too." Quarko seem impervious to justifiable criticism. He continues. "You must admit that I am a quintessential quark."

Ruah is blunt. "I would define you as a quizzical, queasy, and quaky quark. You have some good qualities: your intelligence, your humor, and your loyalty to our mission. Your sowing of the seeds of disunity among the members of the atom families drains your loyalty of sincerity." She concludes, "Could you again pledge a more wholehearted loyalty to our mission?"

Quarko is hesitant. It means swallowing some of his huge pride. Quarkie and Quarkoff center their attention upon him. Finally, he says, "I pledge my renewed loyalty. But don't force me to give up my individuality and some of my quirkiness."

"You have had your freedom, Quarko." Ruah reminds him of the past. She goes on. "We can tolerate some of your humor and quirkiness, but in your loyalty to the creative cause, keep the purpose of our mission in mind."

"You have my pledge, dear Ruah." Quarko seems sincere.

Ruah then redirects the conversation. "In order for the creative process to continue, you need to make Prokye 1's reproductive system more efficient. You have created RNA, with its two functions. First, the RNA combinations of nucleotide bases have written on them the complete body plan of Prokye 1. (We can also call the encoded body plan a template.) Secondly, the encoded RNA bases are transported to the site for the construction of the new prokaryote bacterium. From this template, the atom families build the new prokaryote. They assemble the prescribed strings of amino acids into proteins. The latter are the building blocks for all of the parts and functions of this new being. The RNA-encoded template is then transported back to its special home within the bacterium. So the latter possesses an enshrined copy of its own body plan."

"So we need to do some more experimentation," Quarkoff muses. "We need to keep perfecting our creations to keep the co-creative process on Earth moving forward."

"If we change or modify any of Prokye 1's body plan, we must be sure that these changes are encoded on the RNA template."

Quarkie understands what will be necessary in future cocreating. "Then, when new replications of Prokye 1 are constructed, they will carry the new modifications."

"Both of you understand what needs to be done," Ruah commends and urges them forward. "What about you, Quarko? Any new ideas?"

"Let me outline my suggestion for improvements to Prokye 1's reproduction system," Quarko proudly announces. Quarkie and Quarkoff appear skeptical. "I suggest that we split the two functions now carried out by RNA. Why don't we create a new molecule with a different combination of nucleotide bases upon which the body plan for Prokye 1 will be encoded and stored? This new compound would be located in its special place, its home, within Prokye 1. The present RNA bases would be assigned the tasks of copying the body plan, transporting it to the construction site, and then helping to build the new bacterium."

"Your idea for splitting the two functions is excellent!" Ruah affirms Quarko.

Quarkie is surprised but pleased by Quarko's suggestion. "You have contributed an important suggestion that may help us in this new phase of cocreating."

A short time later, Quarkie, Quarkoff, and Quarko are energetically trying to perfect a new design. With the help of the other atom families, they have tried different combinations of nucleotide bases without success.

Quarko makes a further suggestion. "This is not much of a change. What if we substituted in the sugar group, of which we are a part, a hydrogen atom in place of the present RNA oxygen-hydrogen combination. In other words, we would disconnect the latter from one of our carbon sites and replace it with a simple hydrogen atom. I think this might work to create something new."

"Well, let us see just how intelligent you really are, queasy Quarko." Reluctantly, Quarkoff approves Quarko's suggestion. He seems to hope that his hunch will turn out to be wrong. "Let us try it."

"I am willing to give you a chance, my friend." Quarkie acquiesces outwardly but is dubious on the inside. She makes an interesting observation. "In working with the five nucleotide bases, I notice that the T-base is not being utilized by the RNA molecule. Why don't we substitute the T-base for the U-base that we have utilized in building the combinations of bases to create RNA. (I remember from my biological studies that the U-base is uracil and the T-base is thymine.)

They begin to translate their idea into action. Quarko requests the oxygen-hydrogen pair to detach themselves from their carbon binding site. The duo departs. He hails a nearby hydrogen atom. "My friend, would you be so kind as to become a part of our new molecular group? You just need to bond with our nearby vacant carbon site."

"Dame Destiny has summoned me!" the atom responds. "I am puffed up with princely pride to accept your gracious invitation."

"It is Photie!" Quarkie yells. "I would know his lordly speech and manner anywhere."

"Where have you been, dear friend Photie?" Quarkoff is filled with joy. "We have missed you very much. A short time ago, we lost our two friends, Nitro and Hydrojean, who became members of a new prokaryote bacterium. You remember them."

"Oh, I have a positive momentous memory of them," Photie proclaims. "Homey hospitality they always showed me." He collects himself to answer their question. "I was carried on the wild winds of the ambling atmosphere landing near this serendipitous seashore. Suddenly, I came to the wondrous wall of a bulbous bubble. Famous Fate drew me through one of its hollow holes. I drifted through this sumptuous aqueous sphere till Dame Destiny brought me to your handsome home."

"Excuse me, fatuous Photie." Quarko is a bit miffed. "I am a male quark, not a dame."

Quarkie and Quarkoff can hardly contain their amusement. But they remain quiet.

"Forlorn forgiveness and piteous pardon I beg of you, oh friend Quarko." Photie is gentle in his request for Quarko's absolution. "I mean no dangerous dishonor to descend on you." Photie is very repentant.

"Welcome, beloved Photie." Ruah meets this old friend, radiating love toward him. "You are now a part of very recent new creation, the first prokaryote bacterium, Prokye 1."

"Ruah, privileged I am to be a peerless part of this new crafty creature," Photie responds. "What friendly function do I serve in relation to this beauteous being?"

"Quarko's idea of substituting you, Photie, a hydrogen atom, for an oxygen-hydrogen pair of atoms is especially brilliant." Ruah lavishes praise on quintessential Quarko. "This new substance is identical to RNA, except for your presence, Photie."

"I accept your precious praise, dear Ruah." Quarko is exultant. "My idea of changing one ingredient of RNA was a brilliant innovation. I am sure this new substance has a lofty purpose in our cocreative process. I dislike admitting it, but I heard the sonorous sounds of love energy's lilting music as this new being was born." Quarko is still basking in glory. "What name do you give to my creation?"

Quarkoff and Quarkie are seething with anger at Quarko's greed in claiming the full credit for this new creation. Putting her strong feelings aside, Quarkie says, "Ruah, before you name this new creation, we need to implement my suggestion. In building the new combinations of nucleotide bases that will be encoded on Prokye 1's body plan, we will need to utilize the T-base rather than the U-base with the other three bases that this new molecule and RNA will use in common." Quarko and Quarkoff agree to this suggestion. Together, they construct a new T-base molecule that exactly encodes Prokye 1's body plan. Quarkie concludes by addressing Ruah. "Now you can do the naming?"

"We will call this new creation DNA," Ruah announces with firmness. "Its name denotes that it lacks the oxygen-hydrogen pair but is still made up of the three groups: phosphate, sugar, and nitrate." (I recall from my biological studies that DNA's full name is deoxyribose-nucleic acid.) "In addition, the new DNA molecule's nitrogenous group will use the three nucleotide bases also employed by the RNA molecule, but in combination with a T-base rather than RNA's U-base." Ruah then denotes its function within Prokye 1. "No longer will RNA carry the template of Prokye 1's body plan. The DNA template encoded with its complete body plan will live in its special home inside of Prokye 1. The RNA molecule will become the encoder of the body plan, the messenger for the plan's transfer to the construction site and the builder of the new bacterial replica. If there are changes in the body plan, these modifications will be transferred by RNA and encoded on the DNA template."

Quarkie, Quarko, and Quarkoff implement these suggestions. With the atom community, they construct the new DNA molecule containing Prokye 1's complete body plan and transfer it to its special domicile within the interior waters of Prokye 1.

Quarko is bursting with pride. "I feel vindicated from previous criticism. My suggestions are earth-creating. Wow!"

Quarkoff has had enough of Quarko's egotistical attitude. "Do not claim full credit for this new creation! The creation of DNA was our joint accomplishment."

"You are correct in your assessment, Quarkoff." Ruah agrees with him. "You have overstated your part in this cocreation of DNA, Quarko. All of the five atom family members and other atom families contributed to this new creation. I salute you all."

"Thank you, Ruah. It was our joint efforts that cocreated DNA." Quarkie is grateful. "You have always emphasized that we are a community of beings. When we are successful in pushing the Earth forward in its development, we can all share the credit."

"Maybe I did slightly overstate my importance." Quarko is feeling embarrassed.

"Slightly overstated?" Quarkoff bellows. "Grossly overstated is my estimate of your egoistic claim. Thank you, Ruah, for stating the truth."

"Enough of this bickering." Ruah calls a halt to the quarks' interplay. "You have to move ahead to new creative ventures."

"Where do we go from here, Ruah?" Quarkie is anxious to move forward.

"As I told you in the first moment of your existence during the first second of the birthing of the universe, Quarkie, that is for you and your friends to decide," Ruah encourages them.

Quarkoff sighs. "As a community, we have accomplished another mission on our long cocreative journey!"

Chapter 28
Nitro and Hydrojean Meet a New Friend

"Nitro, will we ever see Quarkie, Quarko, and Quarkoff again?" Hydrojean is depressed. "I even miss quirky Quarko, who made a habit of infuriating me."

"I am feeling very lonely and lost also." Nitro identifies with Hydrojean. "But we must remember what Ruah and Sophia have always taught us—namely, to carry out our cocreative mission. So we have to make the best of our situation."

"That is right, Nitro." Hydrojean is beginning to move away from self-pity to accept their plight and move into the future.

The time is October 12, 3.3 bya.

"We were successful in helping Prokye 1 to make a mirror image of himself." Nitro feel a sense of accomplishment. "Here we are inside Prokye 2's new bubble wall, living in its semi-fluid environment."

"I am going to introduce myself to one of the members of the atom families near me." Hydrojean looks to an atom who is part of the wall which delineates Prokye 2's outer boundary. She greets him and begins a conversation. "Dear neighbor, my name is Hydrojean, and I am a member of the hydrogen family. I would like to get to know you."

"I am very happy to meet you, Hydrojean. My name is Phoson. I am a member of the phosphorus family." Phoson is happy to meet her.

Nitro decides to introduce himself. "Phoson, I am Nitro, Hydrojean's friend. I am a member of the nitrogen family."

"I rejoice in meeting you also, Nitro," Phoson responds.

"Where did you live before you became a member of Prokye 2's crew?" Nitro inquires.

"With the energy of love that radiated from Ruah, I was present at the water's edge at the creation of planet Earth. Amidst tumultuous thunder and lightning, the scalding sea, radiation from the Sun, and the touch of Ruah's love energy, the members of the hydrogen, oxygen, carbon, nitrogen, and my phosphorus family, including myself, were laboring to cocreate a new type of being."

Hydrojean interrupts Phoson. "You know Ruah?" She is shocked that Ruah is working in other parts of the Earth.

"Yes, I have known Ruah, along with Elohim, Sophia, and Dabar, since a time close to the beginning of the birth of our universe." Phoson confirms his acquaintance. "Elohim gave me my name."

"So you know all of the Divine Beings. We know them also, Phoson. That gives us common friendships. I am beginning to feel less lost and lonely." Hydrojean is relaxing with this new knowledge. "Go on with your story."

"We were trying various combinations of molecules to form long trains of compounds," Phoson continues. "Finally, we were successful. Within the bubble wall of which I was a part, we were able to construct a network of elements to form a new creature. Through our efforts and Ruah's indispensable vivification, this new being began to live on its own as an individual. This new being that we cocreated Ruah called a prokaryote bacterium." He pauses briefly and continues his story.

"Sophia challenged us to work to cocreate a combination of molecules that would allow our bacterial being to make a mirror image of itself. In this process, our five atom families were able to produce a compound called RNA, made up of phosphate, sugar, and nitrate groups. The latter nitrate group was different. It was made up of five nucleotide bases. They were able to be arrayed in different combinations of bases upon which the body plan could be encoded. The RNA-encoded bases were able to be transferred to a construction site, where the building of a bacterium replica could be accomplished. Working from the RNA design plan, we built strings of amino acids, present in the semi-fluid environment, into proteins, which became the building blocks of all the parts and functions of the bacterium replica. Then we transferred this RNA body plan to a special place within our bacterial host. This became the home of RNA. Thus, we were successful in building and launching into the environment outside our first prokaryote bacterium replica."

"In the same way, Sophia also challenged the members of our community living within our bacterium, Prokye 1." Nitro shares his common experience with Phoson. "So our community created an RNA-based body plan, and we were able to create a reproduction of our host, Prokye 1. We ended up being enlisted to become a member of the crew of this new bacterium, Prokye 2."

Phoson takes up his story again. "After creating RNA, I was replaced by another family member in the bubble wall of our prokaryote bacterium. As an unneeded atom, I ended up being discharged through an opening in the walls of my bacterial home into the sultry waters outside." Phoson continues. "I drifted in these sweltering waters for a time and was able to enter and find a temporary home inside still another bacterial creature. The latter was a bacterium that had been replicated from an existing prokaryote bacterium. Within the walls of my new home, our five atom families were working on still another project at Sophia's request. We were trying to improve its structure. Our hard work, through trial and error, created a new molecule, DNA, from RNA. To accomplish this, our community replaced an oxygen-hydrogen combination at a bonding site of the sugar group with a single hydrogen atom.

In addition, like RNA, a DNA molecule possessed five different nucleotide bases. One of our members noticed that one of the five bases, the T-base, was not being utilized by the RNA molecule. So we substituted the T-base for the U-base in developing the new DNA molecule. RNA and DNA both employ three bases in common: the G-base, C-base, and A-base. (In my biological studies, I recall that these three nucleotide bases are guanine, cytosine and adenine.) Using these four bases, we were able to build various combinations of these nucleotide bases upon which the entire design for the prokaryote bacterium could be encoded. So DNA assumed one of RNA's functions as the custodian of the body plan. RNA kept its other functions: the copying of the design from the DNA template, transferring the copy of the design to the construction site, and then building the strings of amino acids to make the proteins, the building materials used to fabricate all the parts and functions of the new bacterial replication."

Hydrojean is surprised to learn of this new development of DNA. "We developed RNA but have never heard of the new creation, DNA."

"So your creation of Prokye 1's RNA was able to meet Sophia's challenge." Phoson understands that Nitro and Hydrojean's community was also challenged. "Our community of atoms was able to replicate our host bacterium numerous times before we developed DNA." Phoson pauses. He continues. "Once again, I was discharged as waste from my second bacterial home. I ended up entering into the semi-fluid interior of your friend, Prokye 1. Your atom community used me in the creation of Prokye 2, presently our joint home."

Suddenly, Ruah comes into their midst. "Greetings, dear Phoson, Nitro, and Hydrojean. I can see that you have already introduced yourselves."

"Hydrojean took the initiative to bring us together." Nitro is grateful to her. He has a question to ask her. "I now understand that you, Ruah, have been in touch with all the watery edges of the earth and that you have worked with the five atoms families to proliferate the cocreation of prokaryote bacteria. Phoson told us of Sophia's challenge to his former communities. Now that we have met Sophia's challenge, what is our next task?"

"I cannot give you any specific directions for the future." Ruah is honest. "Sophia, Dabar, and I have consistently stressed the need for you to take the initiative. You are cocreators. In general, I can say that you need to create DNA, a task already accomplished by many other prokaryote bacteria crews, including your friends living in Prokye 1. Phoson, you know the formula for creating the DNA molecule. Lead Nitro and Hydrojean in developing DNA for Prokye 2 to take the load off of RNA." She reminds them, "Remember that if you make modifications to the body plan of your prokaryote bacterium host, you need to enlist RNA to carry these changes to the DNA's home and encode them upon its design plan."

"I am ready to help Nitro, Hydrojean, and the other atom families in the task of constructing the DNA molecule and encoded bases. I know what is necessary," Phoson responds to Ruah's request.

Led by Phoson, Nitro, and Hydrojean, the five atom families build the DNA molecule and its encoded nucleotide bases. Then they turn their attention back to Ruah.

Ruah addresses them. "Continue to work to cocreate a continuing diversity of creatures. Prokye 1, Prokye 2, and their vast bacterial family are only the small beginning of what might be accomplished through our cocreative activities." Ruah pauses briefly. She finally asks, "What do you think you can do to push forward creativity on planet Earth?"

Phoson has a suggestion. "While I was out in these coastal waters, I observed numberless prokaryote bacteria. They were constrained from moving around in an effective way. They need a way to move themselves from place to place more efficiently."

"Phoson, that is an excellent observation," Ruah commends him. "Within the bubble walls of Prokye 2, you, Hydrojean, Nitro, and the five atom be accomplished through our cocreative activities of the bacterial family to move itself from place to place."

"I am ready to help you, Phoson, in this task," Hydrojean volunteers.

"Count me in also, my friend." Nitro is up to the efforts involved in this new goal.

"Keep in mind, Phoson, Hydrojean, and Nitro, that the work of perfecting such a mechanism of locomotion may take a long time," Ruah cautions them. "There will be many trials and frustrations. You need to persevere."

"Ruah, the communities in which I have lived up to the present have always worked together to move the cocreation of planet Earth. I have no doubt that this community of our five atom families within Prokye 2 will give the same sterling effort." Phoson is full of enthusiasm.

Chapter 29
Photie and Phosie Volunteer
for a Creative Journey

"Dame Destiny brought me to your charismatic community, Quarkie, Quarko, and Quarkoff." Photie is philosophizing about his life. "She has more mystical magic to enshroud our beauteous beings and lead us to new daunting destinies."

"You are a master of alliteration, dear Photie." Quarko is complimentary for a change. "Your vital virtuosity challenges my immense intelligence."

"Our challenge and mission given by the Divine Beings will lead us to 'daunting destinies,' I feel sure, oh faithful Photie!" Quarkie confirms Photie's intuition.

The time is October 16, 3.2 bya.

In this communing moment, another familiar voice is heard by way of greeting. "Hello, my friends." It is Phosie. "I am also a member of Prokye 1's community. Presently, my place is living within the phosphate group that makes up Prokye 1's RNA." Phosie is happy to be reunited. She addresses Photie. "I saw you enter Prokye 1 through the opening in our bubble wall. Initially, I did not recognize you in your new hydrogen guise. I was so used to you being a fleeting package of yellow energy."

Querulous Quarko responds to Phosie. "First you were a part of the wall that separates us from the society outside of ourselves.

Then you had to be released from that position. Because you were incompetent. Now you are employed by the phosphate group in Prokye 1's RNA. At least you are good for something."

"Disregard our discourteous quark mate, Phosie." Quarkie critiques her mate. "Your work is important to Prokye 1's existence. Your phosphate family is the bubble barrier that defines our bacterial friend, Prokye 1. Your family's openings in these walls allow new atoms and molecules to enter Prokye 1's fluid environment to give us the raw material to continue our cocreative work. In addition, those materials not needed are able to be discharged back into the watery world outside," Quarkie affirms. "Together, we have labored to cocreate the RNA to help Prokye 1's replication."

"Yes, we, members of the phosphorus family, are proud of our joint efforts," Phosie responds. She then adds, "We have much more cocreating to accomplish."

Ruah has been eavesdropping. She says, "Yes, Phosie, much, much cocreating lies ahead. You cannot even dream what amazing accomplishments may lie ahead." She receives their attention. "I do not know exactly what they will be. I trust that you are still eager to keep cocreating and diversifying this new life on planet Earth."

"Momentous moments do lie ahead." Photie is confident. "Fascinating Fate will lead us to amazing accomplishments. Happy hope impels us to our gaudy goals."

"Wordy wisdom, fancy Photie." Quarko mocks Photie. "If wild words can cocreate, we will experience a fantastic future of devilish diversity." He smirks as he ends.

Photie is oblivious of the mockery. "Oh thank you, thank you, Quarko. Your caring compliment brings me pleasant pride."

"The cocreative mission must continue, my friends, if 'fascinating fate' is to end in 'amazing accomplishments,' to quote friend Photie." Ruah is responding to Photie's prophecy. The latter is smiling as his words are repeated by Ruah. "What is the next step of cocreation? It is up to you to continue carrying out your important mission."

There is a brief time of silence as the group savors and absorbs Ruah's question. Finally, Photie breaks the silence. "Dame Destiny needs an intriguing initiator. I volunteer to be an ambitious apostle of courageous cocreation." He addresses Ruah directly. "Allow daring DNA and rambunctious RNA to include me in the next rapturous replication of prodigious Prokye 1."

"That's the vibrant spirit necessary to push forth the diversification of planet Earth, dear Photie." Ruah commends him. Photie is pleased and ready.

"Since we are a part of Prokye 1's DNA, it up to us." Quarkie begins to support the momentum begun by Photie. "We will summon the four RNA nucleotide bases to plug into our DNA's encoded body plan to make a copy. Then the RNA bases will become the messenger, bringing our design plan to the construction site. Then the atom community can begin to translate this plan into a mirror image of Prokye 1, a new prokaryote bacterium." She pauses and looks intently at Photie. "Be ready to be incorporated into this new being, friend Photie."

"Dazzling destiny has summoned me, Quarkie." Photie affirms his readiness. "I await resilient RNA's command."

I watch the RNA bases do their copying, messenger, and construction work in conjunction with the atom community. Photie, as a hydrogen atom, is incorporated in the DNA bases of this new being.

As Photie leaves, the quark trio bid him a fond farewell. "Goodbye, dear precious friend, Photie!" Quarkie is weepy.

"We hope to see you in the future." Quarkoff is beginning to feel his loss. "'Courageous cocreating,' fond mate."

"May alluring alliteration accompany your 'courageous cocreating,' phonetic Photie." Quarko seems sincere. After a pause, he asks Ruah, "What do we call this new creature?"

"Photie is a member of Prokye 3." Ruah names it. With this naming, Prokye 3 departs through an opening in Prokye 1's exterior wall.

Upon Photie's departure, Quarko issues a new challenge. "Who is the next volunteer for 'courageous cocreation?'"

Momentarily, there is silence. Suddenly, a new volunteer comes forth. "I am ready! Send me." It's the voice of Phosie. "Recently, my place in RNA's phosphate group was taken by another family member. So I became a waste atom within Prokye 1."

"Are you sure that you want to leave us?" Quarkoff wants reassurance.

"Photie inspired me." Phosie is enthusiastic. "'Courageous cocreation' is necessary for the success of our challenge and mission. Although the future is unknown, I am ready to move into fascinating space-time."

"We will summon DNA and RNA to begin the replication process." Quarkoff initiates the call.

Almost immediately within the DNA's home, and later in the semi-fluid construction side, RNA begins the replication process. Phosie is incorporated as a member of the bubble wall of this new prokaryote bacterium. Soon, the replication is complete. This new prokaryote bacterium with Phosie on board is about to depart.

Ruah names this new bacterium. "Phosie is part of Prokye 4. As with Prokye 3 and Photie, their mission is to cocreate and diversify. This will be important for the future development of planet Earth."

The quark trio bids their gracious farewell to Phosie. "Best wishes and 'courageous cocreating,' friend Phosie."

With Phosie in residence, Prokye 4 departs through the bubble wall's opening. This new bacterium takes up its citizenship amidst the fast-growing prokaryote bacteria community on the shores of planet Earth.

Quarkie is again feeling a deep sense of grief and loss. First, it was Hydrojean and Nitro. Now Photie and Phosie. "I am depressed again. All of our intimate friends have departed. I feel lonely like I experienced in the first moment of my creation."

"But you have us." Quarkoff tries to reassure her.

"Quintessential Quarko stands at your service, madam." Quarko tries to brighten her spirits.

Ruah urges them forward. "I am afraid you will not be able to stay in this morose spirit for long."

"What do you mean, dear Ruah?" Quarkie questions her.

"In the beginning when the universe was created, space-time was created," Ruah begins. "We, the Divine Beings, have lived from all ever. We live always in the present time. We had no beginning; we will have no end."

"So you live in different time than we do," Quarkoff exclaims.

"Yes, your time, Quarkie, Quarko, and Quarkoff, includes a past, present, and future." Ruah differentiates the two types of time. "I mention this because I want to take you on a journey back to the past."

"Back to the past! And a journey!" Quarkoff comments. "We will travel back into space-time?"

"Yes, you will travel back into space-time. Presently, it is October 19, 3.0 bya," Ruah affirms Quarkoff. "You live in a microcosm of the universe and planet Earth. This allows you only to be present to your minute surroundings. You are unable to see what is happening in the macrocosm of planet Earth."

Quarkie comments, "Over a billion and a half years ago, you took us on a tour of the Milky Way galaxy. Will this journey be like that?"

"It will be something like that trip, with a difference," Ruah explains. "During your galaxy journey, you were living in present time. In this travel upon which we will embark, you will be going back to past time."

"Why would you want to take us on such a journey?" Quarko is inquisitive.

"I would like to show you how the earth came into existence and how it has developed during this past billion years up till this present time." Ruah explains the reason for the trip. "You need to see how all of the ninety-two atom families, including the six families involved in biological life, are working to cocreate planet Earth. In addition, we will look at the whole solar system." He pauses. Then he ends his explanation. "Get ready to leave with me!"

Quarkie responds for the trio. "We're excited, Ruah. Lead us back to the past!"

Chapter 30
Back to the Past: The Formation of the Solar Planets

"I have been present to this scene before," Quarkie remarks.

"You were here at the time of the 'provoking detail' prophesied by Sophia. The time is early September 10, 10.4 byich." Ruah reminds her of the inaugural event that set off this new stage of creativity.

"The explosion of the supernova!" Quarkoff names this "provoking detail" with pride.

"A very, very memorable event," Quarko recalls. "Our trinitarian trio was crushed together in deep intimacy. I again feared we would end up an omelet. After the gigantic wave of energy passed, we returned to normal." Quarko is amazed to come back to this scene.

"Ruah, as I viewed this cataclysm from a distance, I remember that from the part of the vast nebula where we were situated, I saw a new star being spun out of the hydrogen and helium gases." Quarkie describes what she had seen. "The supernova's stupendous yellow-red blast sent terrifying shock waves and shot multitudes of space debris in all directions. Already, this stellar nursery was full of dust and gases. The supernova energized the spinning action of this budding sun."

On this journey into the past, this proto-sun comes into view. That section of the kaleidoscopic nebula is being swept clean of the nearby golden helium and orange hydrogen by its spinning action. This past event is being speeded up to whirlwind speed. The quark trio is taken by Ruah inside the proto-sun, close to the boundary where the immense pressure is converting the orange hydrogen atoms into the golden helium 4 nuclei. The latter continues to fuel the furnace of this birthing sun. They watch as the temperature reaches a fiery 18 million degrees. Suddenly, a white-hot flash ignites the nuclear furnace's core. The quarks are caught up in the yellow, flaming tornadolike radiation that shoots out in all directions from the torrid center of this vibrant, newborn sun, newly ignited.

"Wow!" shouts Quarko in the midst of tumultuous wind that shoots them through the interior of the Sun and out into space. Intrastellar expanse is a crowded chaos of awesome gray asteroids of various monstrous and smaller sizes, multitudes of flashing ice-loaded turquoise comets, and swarms of silver-gray meteorites.

"Save us, Ruah!" Quarkie is filled with fear.

"I would rather not be back to the past." Quarkoff feels threatened with imminent danger.

"I like this rapid space ride, Ruah!" Fearless Quarko, like his mates, is beginning to slow down slightly as he continues to drift with his quark mates through that part of the nebula out of which this sun has been spun. "I have never been this exhilarated and energized. I could ride on forever."

This rapid journey into nebular space has caused them to be driven to the outer limits of the solar system of planets being created by this supernova explosion.

Overcoming her initial fear, Quarkie observes, "What a stupendous space ride. Now that we have come to a stop and I look out from our solar system, I can see many, many other suns that are being spun out of the nebular gases in this vast cloud. That supernova really energized this stellar nursery. The hydrogen and helium families are very busy creating this part of the universe, Ruah."

"You see creativity happening all around you, my friends. You are observing this creative activity only in this tiny part of the outer arm of the Milky Way galaxy." Ruah responds to Quarkie's observations. She continues. "I have told you all along that you are not alone in cocreating the universe. All of the ninety-two atom families are engaged in this birthing process in all of the many billions of galaxies in this universe. I am a source of vibrant love energy for their creative projects." Ruah pauses and then goes on. "Now look back at your newborn sun that is in the center of this solar system of which you are a citizen. What do you see?"

"I see about nine planetesimals being created, including our Earth, circling in different orbits but traveling in the same direction around our brightening yellow-white sun." Quarkoff is quick to answer. "They are of different sizes and at varying distances from the sun."

"The solar system is not shaped like a round bubble, but like a narrow, oval sphere." Quarkie makes this observation. "Most of the planets orbit on a plane around the sun."

"Do you notice anything else?" Ruah is testing their skills of observation.

Recently, Quarko has been noticeably quiet. He now enters the dialogue. "Are you testing us, Ruah?" Confidence emanates from his person and voice. He takes up the challenge. "We are at the far edge of our solar system. This area of the solar system is cold, very cold. The farther we are away from our luminous sun and its intense radiation, the colder it is." Quarko looks at Ruah intently. "Have I sufficiently answered your question?"

"You are awake, Quarko, and observant!" Ruah commends him.

"You do not have to be so boastful, Quarko." Quarkie is tired of his triumphant attitude. "One thing that I notice is that these five outer planets are attracting gases, frozen dust particles, and icy iron rock. Ruah, your weak energy force, gravity, is helping them sweep up all the lighter gases and some of the heavier atoms in their vicinity."

Ruah takes them inside one of the gaseous planets. "Can you see that gravity is attracting the heavier atom families into its core? The lighter gaseous elements are being frozen and are gathering toward the outer portions of the planetesimal. The same process is going on in all of the other five outer planets. We will have to wait and see what the finished planet will look like."

Ruah guides the quark trio past the five outer gaseous planets into the inner belt of four planets. These bodies are in various stages of formation.

"We have to be careful here," Quarko cautions. "These planets are all yellowish red in color and molten. You need a traffic director, Ruah, to guide the huge, gray asteroids and classy comets with their long, azure, icy tails into softer landings on these four molten planets. The atmosphere around each one is filled with dark, pungent clouds generated by the fiery chaos raging on their red-hot molten surfaces. No sooner does each planet begin to settle down when additional asteroids, meteorites, and comets pummel their fiery, melted surfaces." Quarko pauses. "Ruah, you are infinitely wise." Sarcasm begins to infect his words. "How will we be able to cocreate new beings, with your assistance, of course, on the third planet from the Sun, our Earth?"

"You are so smart, Quarko." Quarkoff readies to blast him with sharp words. "Remember, stupid, Ruah has taken us on a journey into past time. We have already cocreated on our planet Earth in the present. Have you forgotten, you queasy quark?"

Suddenly, their gaze is drawn toward the third planet from the Sun, the Earth. An ominous menace approaches this molten proto-planet. This fateful asteroid is mammoth in size, possibly a quarter the size of this accreting, fiery ball. If it strikes the molten Earth directly, it will annihilate itself and this newly forming planet.

"Ruah, disaster is pending!" Quarkoff shouts a warning. "There's going to be a cataclysmic collision." This gigantic mass continues to target chaotic, red-hot, boiling planet Earth.

Quarkie is frozen in dire anticipation of a bombastic impact. "Look, a colossal crash is close at hand."

"It is veering off course, Quarkie." Quarko is excited. "It will not be a solid hit. In fact, it may miss striking the earth altogether."

This huge, gray mass begins its final descent. Quarko is correct in his observation. It will either miss this fiery Earth or strike it with a glancing blow.

"This asteroid has just hit the edge of the scalding planet." Quarkoff is transfixed by the event. "It is sweeping up some of the red-hot molten magma from the earth. It is bouncing off the earth and moving out into space."

"This gray mass is being consumed by an intense fire set by the red, boiling magma that it swept up in its collision." Quarko watches this torrid mass move further into space. "As it travels out into the area near the Earth, it is slowing down." Quarko pauses thoughtfully. He then asks Ruah a question. "Why is this mass slowing down as it departs the earth's surface?"

Ruah responds to him. "Quarko, this great molten mass is slowing down because the weak force of love, gravity, is asserting its power. The Earth, being a larger body, is able to attract this smaller planetesimal by this love force of gravity and move it into an orbit around the planet Earth. Do you remember that Elohim gave the name to a smaller planet that orbits a larger planet?"

"Yes, I do remember." Quarkie is the first to recall this naming event. "Elohim called it Moon. So being back in the past has allowed us to witness the birth of this Moon planetesimal. It was an exciting birth." During a lull in the conversation, she muses and continues. "When we move forward in this journey, still in the past, we should see how the Moon has developed."

"That's correct, Quarkie." Ruah affirms her insight. "We will see how it has evolved with the help of the atom families that populate it." She adds an important comment. "The Moon is especially important in helping the Earth to keep its tilt toward the Sun. If the planet's tilt was erratic, creative activity would come to an end." Ruah continues. "We need to move ahead now in this past time to observe the culmination of the entire creative action taking place in the four inner and five outer planets."

"So we are still in the past, but we are going to move a few days in universe history to see how each of these planets evolves." Quarkie interprets Ruah's statement.

"Yes, Quarkie, we are going to move forward in the past to September 25, 4.0 bya. It's the time closer to your work in cocreating prokaryote bacteria." Ruah confirms her interpretation.

"Take us forward in the past, dear Ruah." The quark trio is united in moving forward.

Chapter 31
Back to the Past: A Visit to the Four Rock Planets

"Once you entered the microcosm of the planet Earth to do your creative work, my friends, you lost contact with the macrocosm of the universe, especially your solar system and its nine planets." Ruah is reminding them why she is taking them on this journey back to the past. "As I told you when we began this journey, I wanted you to know what the other families of atoms were doing while you have been busy creating biological life."

The time is September 23, 4.0 bya. Ruah is conducting the quark trio again on a journey through the Sun and its solar system of planets.

Ruah had journeyed swiftly with Quarkie, Quarkoff, and Quarko through the past time, from September 10 to September 23.

"You can certainly cover past time in a harried hurry, Ruah." Quintessential Quarko is again the noble quipster. "I was hoping that the rapidity of our journey would have distanced me from my forgettable quark mates."

"Thanks for your stunning support of us as your shipmates." Quarkoff sounds bitter.

Quarkie is more concerned with her view of the solar system and its nine planets. She is presently turned toward the Sun and making some observations. "Our scintillating Sun is gigantic in size since we observed its origin. It has been growing in brightness and intensity. At a long distance, it looks perfectly round and its surface luminous and fiery yellow. Close to our Sun, I can see enormous yellow-white flames exploding from the Sun's surface. Huge, energetic blasts of radiation are been sent into the intrasolar space. In addition, I see multitudinous swarms of yellow photons, our friends in the yellow packages of energy family, being shot in all directions from our Sun, giving glorious light and warmth to planets and space. I have never taken time to observe these characteristics in the previous suns of which I and my quark mates have been a part."

"We did not have time to observe the suns. We were usually being launched and rocketed into expansive space or scrunched together like our three helium 4 nuclei were when we created the first carbon atom." I waited for a punch line. Quarko fooled me by continuing to follow-up on Quarkie's observation. "When that last mountainous, yellow-red flame leapt out from the surface of our Sun, the blast of radiation almost flattened my frame. What a boisterous, battering blast!"

Ruah takes up the conversation at this point. "Yes, your Sun is powerful. Its ever-flowing photonic energy and fiery flaring affects all of the nine planets, but it will be especially important for the third planet distant from it, your planet Earth."

Quarkie has turned away from the Sun. With her quark mates and Ruah, they have traveled swiftly above the plane of the Sun and its planets. She notes, "Ruah, the solar system is similar in structure to our Milky Way galaxy, yet different. Our Sun is like the core of the Milky Way. The nine planets are like the arms of the galaxy, but are not attached to the core. The four smaller solid planets and the five gas planets are orbiting around our Sun at varying distances from it. With all the planets moving in the same direction, their loping orbits are etched like circling lines on a flat platter." Quarkie concludes, "The seventh planet out from the Sun seems to be lying on its side. Strange!"

"We will get a closer look at that planet as we travel out from our present location near the Sun." Ruah responds to Quarkie's comment. Then she continues. "Let us move through the solar system and observe each planet."

"The first rock planet looks like it has taken many blows from out in space." Quarkoff is describing what he is observing. "I remember back on September 10 when it was being formed. Very large and smaller gray asteroids and flashing silver-gray meteorites were impacting this already accreting planet." He pauses and asks Ruah, "What caused this planet and the other three rock planets to be on fire and molten?"

Ruah responds readily. "Good question, Quarkoff." The latter feels affirmed and recognized. He exudes pride. "The members of a special family of atoms, the volatile families of atoms like uranium, are very energetic. They fueled this fire with their special love. The heavier, denser atom families, like molten iron and nickel, began descending to the center of each planet. The other, lighter atom families were fused gradually into the violent, red molten magma. As these planets began to cool, they formed a gray crust. Some of the reddened magma were spewed out through many, many cracks in this crust and began to further pave the surfaces of the rock planets."

"Have you given a name to this outpouring of crimson magma through cracks in the crust of the rock planets, Ruah?" Quarkoff is again questioning the ghostly godmother.

"Quarkoff, thank you for asking." Ruah is again complimentary. "Since the magma is so hot, reddish, energetic, and volatile, I will call this manifestation a volcano."

"Dear Quarkoff, from what secret source has all of this intriguing intelligence come?" Quarko is again an active irritant. "Maybe your dull wit has been sharpened and jostled by this celeritous journey into past time."

"Let's ignore the quipster." Quarkie is quick to change the subject. "In the beginning of the four rock planets, the surface was a fiery chaos. Now these planets possess a hardened surface. I do see multitudinous numbers of red, belching volcanoes pouring out their scarlet hot magma, which cools and hardens. I also observe these many very large holes, along with other, smaller indentations." She is pensive for a moment. And then she asks, "What has caused these hollow indentations on the surface of the rock planets?"

"Quarkie, although most of the debris that used to surround the planets has been gobbled up by rock planets, there are still some stray asteroids and meteorites that occasionally are drawn to the rock planets by the love force of gravity. When they impact these rock planets, they create these many holes." Ruah answers her question. The ghostly godmother is happy that Quarkie has sidetracked the ever-mischievous Quarko. She gives a name to these holes. "Let's call these holes craters because of their shape."

"Ruah, could we explore the interior of the first rock planet? It is by far the smallest of the rock planets." Quarko is inquisitive. "When we visited this planet on September 10, it was a small, scarlet, broiling planetesimal. Now it has a crust that is potted with craters. What about the inside of the planet?"

"This first planet from the Sun does have a hard crust," Ruah comments. (I recall from my science course that because this planet circles the Sun, the quickest of all of the nine solar bodies, it was named Mercury, who was the Roman god of speed. It is situated 33 million miles from the Sun. Mercury is a little less than half the size of Earth.) "Let us descend below its brown-gray crust. The many heavier atom families have created a hardened layer underlying the crust. Let me name this layer. I call it a mantle. You will find such a mantle in all of the other three rock planets, although the Earth will possess a different type of mantle. I will explain this when we explore the Earth. Below the mantle are the very heavy atom families, iron and nickel. They form its hard center, which is the thickest of all the planets." Ruah concludes her remarks. "You will find that the second and fourth planet from the Sun will have a similar crust, mantle, and core. Again, your planet Earth will be different."

Ruah and the quark trio now ascend to the surface of this planet. Quarkoff comments, "I notice that this planet is moving in an orbit around our Sun. When we are facing the Sun, the atmosphere is extremely hot. After a period of time, the Sun disappears. We are in the dark and it is very, very cold. What's happening to cause this?"

"You stupid quark!" As usual, it is the quirky quark, Quarko, enflaming the situation. "Do you not realize that as this planet and the other eight make their journey around our Sun, it is spinning like a proto-sun in its nebula nursery?" Quarko is filled with pride. Triumphantly, he adds, "The only difference is that these planets are spinning at a constant speed in their journeys, while the proto-sun continues to increase its rate of spin until ignition of its nuclear furnace. This planet's spinning motion is the slowest of all the planets. When one side of this planet spins away from the Sun, that side is in darkness. Later, this side spins to face the Sun again. What do you think of that explanation, imbecile Quarkoff?"

Quarkoff is livid with anger at his quark mate. "You egotistical, queasy quark."

Ruah tries to halt the conflict. "Why are you so adversarial, Quarko?" Ruah looks toward Quarkoff, who is boiling with anger. "Quarkoff, you are intelligent. Quarko's characterization of you is wrong." The former begins to simmer down after Ruah's compliment and his admonishment of the latter. Ruah goes on. "You are right in your observations, Quarko. When this planet is facing our Sun during its periodic spin, the surface is very, very torrid. When that part of the planet turns away from our Sun, it loses the intense radiation of the Sun and is extremely dark and cold. Soon, the rotation of the planet takes that side back to face our Sun again and it heats up. Like all of the nine planets, it continues to rotate at a constant speed as it completes its annual orbit around its luminous Sun."

"I notice that there do not seem to be any gases that surround the planet and give it an atmosphere like we have on planet Earth," Quarkie observes.

"That's right, Quarkie!" Ruah responds. She is happy that Quarko is quiet. "The solar radiating winds are so strong that they blow most of the gases away. Are there any other questions?"

"I do not have any further questions about the first planet from our Sun." Quarkie then directs the attention in another direction. "Let's journey to visit the second planet from our Sun, the yellow-colored planet. It is much larger than the first planet." (I remember again from my earthly life that this planet is called Venus, after the Roman goddess of love and beauty. It is 67 million miles from the Sun and about the same size as the Earth.)

"Why is this planet yellow in color?" Quarko seems a bit subdued as he makes this query.

"What do you observe as we near this planet?" Ruah directs his question to the quirky one."

"There are clouds of gas that are surrounding the whole planet." Quarko assumes a serious vane for once. "With our incandescent, radiating Sun illuminating this gas, it takes on a yellow hue. I notice, too, that the clouds are being blown around the spinning planet. I do not recognize the type of gas."

"It's carbon dioxide gas." Ruah answers his question. "Your carbon family, along with oxygen family, has created this cloud covering the planet. As we enter into the cloud of gas, it has more of a light gray color."

"I notice, Ruah, that it spins in the opposite direction from all the other planets and at a slower speed than our Earth." Quarkoff is inquisitive.

"That is one of the strange features of this second planet from your Sun," Ruah responds.

"The temperature within this planet seems hotter than the first planet from our Sun." Quarko is still in a serious mood. "Why is this so?"

"Although this planet is farther from our Sun, when the radiating heat of our Sun penetrates this gaseous cloud, the latter will not allow the heat to escape into space." Ruah gives the quark trio an explanation to Quarko's question. "The trapped radiation heats up the surface of this planet." Ruah pauses. She continues. "Again, you know that the interior of this planet is made up of a hard crust, mantle, and core created by many cooperating atom families."

"I notice that there are craters on its surface, but they are of smaller size than on the first planet. Why?" Quarkoff asks.

"When stray asteroids and meteorites come through the carbon dioxide atmosphere, it slows these masses and they make a softer landing. Thus, they create smaller craters. Notice the mountains, hills, and valleys created by the earlier volcanoes that at one time flourished in the early times of this planet." Ruah gives him an explanation. Since the quark trio becomes silent, Ruah proposes the following. "Let's move on to the third planet from the Sun, our Earth."

Ruah and the quark trio's journey into past time is moving forward. The time is October 1, 3.8 bya. The volatile, molten surface of the Earth of September 10 has calmed. Meteors and asteroids occasionally impact the Earth, disrupting its more serene condition. Islands of gray crust are forming upon its surface. In addition, numberless volcanoes still convulse from the multitudinous cracks in these islands of crust. They belch forth their load of scarlet magma from the molten mantle that lies underneath the crust. At this time, the watery oceans make up 90 percent of the planet's surface, the islands of crust 10 percent. The volcanoes also shoot water from the red-hot mantle to fill the vast, rocky water basins. These vast, depressed areas have been paved by the black basaltic rock recently spewed out by underwater volcanoes. Torrents of water fall from the atmosphere to fill the ebony rock basins. (The Earth is located, I recall, 93 million miles from the Sun. It takes the Earth 365 days to circle orbit the Sun.)

As the sojourners approach the third spinning planet from their Sun, Quarkie makes this observation, "The clouds covering the Earth seem darker and gloomier than in the previous planet. Is this atmosphere the same as the second planet from our Sun?"

"Carbon dioxide makes up the greater portion of the gases in the air surrounding the Earth." Ruah is quick to respond. "There are other gases created by the various atom families, molecules of hydrogen sulfide, methane, ammonia, nitrogen, water vapor, traces of formaldehyde, and hydrogen cyanide."

"Well, that latter compound is the pungent compound that our carbon atom created when we joined one member of the nitrogen family." Quarko demonstrates quick recognition of this past event. "Notorious Nitro created the molecule, ammonia, with three members of the hydrogen family." He smiles and continues. "Tempestuous Hydrojean formed the molecule, formidable formaldehyde, with other members of the hydrogen family, one member of the carbon family, and one of the oxygen family." As an afterthought, he says, "So we were not unique in creating these molecules, Ruah."

Quarkoff seizes this opportunity to spar with Quarko. "How slow you are to understand, oh high and mighty Quarko, that Ruah and her energy of love are working throughout all of the atom families to help us cocreate."

"Now, now, Quarkoff. No lording it over quintessential Quarko." Ruah steps in to stop any further sparring. "You are all aware that we, the Divine Beings, are present in all areas of this vast universe helping all of the ninety-two atom families to cocreate the universe." Ruah is attempting to defuse any conflict between the two quarks. "Let's explore the depths of the Earth."

"Yes, let's do this." Quarkie seizes on Ruah's offer. "You mentioned that the core of our planet Earth is different from the other three rock planets closest to our Sun."

Ruah guides them to the depth of Earth to its core. "The cores of the other planets were solid with the denser elements of iron and nickel deposited there. In the Earth, at its depth, there is a small, solid core of these heavy metals, which sank to the core due to their higher density. Surrounding the tiny solid core is an outer molten core of these boiling metals. The volatile atom families, uranium, thorium, and potassium, are decaying and creating immense heat to do the melting. Above this is an inner mantle of softer, heated compounds of some less dense metals, sulfides, and silicates. These materials also inhabit the upper mantle near the island crusts of the Earth's surface."

"How did all of this material arrive on the Earth?" Quarkie asks.

"You remember that scene of the asteroids, meteorites, and comets impacting the Earth," Ruah begins her reply. "They carried all these materials and their atom family members to the Earth. It took time for the denser metals to migrate to the core and be fired by the volatile decaying elements. The less dense elements like metals, sulfides, and silicates stayed above the core heated by the volatile elements."

"So there is a stabilizing of the interior of the Earth." Quarko starts to give a serious inquiry. "The volcanoes on crusted islands and at the bottoms of the deep, watery basins must be fueled by the red-hot mantle in the interior of the Earth."

"You are quite observant, Quarko." Ruah quickly adds, "Do not let this compliment go to your head and cause an outburst." Quarko absorbs this comment without a blustery response. "There is a constant ebb and flow between the inner and outer mantle. When the icy comets came to Earth, they were laden with water. Much of this was deposited in the mantle underneath the Earth's crust. The boiling magma is looking for a way out. When it finds a fissure in the crust, it begins to expel its reddish, molten magma, water vapors, dark clouds of ash, and other gases."

"Do the torrential rains falling now on the Earth come from this same source, Ruah?" Quarkie is inquisitive.

"Yes, Quarkie. As the upper atmosphere cools, the water vapors condense and fall as rain, carrying with them many types of molecules, which settle in the water basins." Ruah affirms her reasoning.

"So that's how all of us, new molecules of hydrogen cyanide, hydrogen sulfide, ammonia, and formaldehyde, were ultimately carried to Earth and were deposited near the edges of the basins of water." Quarko makes a new deduction.

"That's correct, Quarko." Ruah affirms his insight. "The molecules of water created by hydrogen and oxygen families of atoms were your transportation to the crust of the Earth." Ruah muses. "You spoke of the basins of waters. Let me give them proper names. Let us call the large basins of water oceans, the smaller basins seas, and the tiny basins lakes or ponds. So we can say that you were transported to the seashore, my quark friends, by the molecules of water that fell and are falling upon the surface of the Earth."

"It's good that you give names to these places, as I strain to find words to express experiences I have had on this Earth. Thank you." Quarko is bubbling with politeness and joy.

"Keep in mind that the inner and outer mantles are volatile. Some of the materials from the upper mantle, including water that flows down through the cracks in the crust, descend to the inner mantle. Then elements and compounds from the inner mantle float up to the upper mantle. If there is water present, it makes the upper mantle much more erratic and volatile. This new magma, made up of light atoms from the depths, is blown out onto the crusted islands of the Earth, paving its surface. Ash, gases, and water vapors are blasted into the atmosphere, replenishing those lost by being carried to Earth by the rain water. There is a continuous cycle of diminishment and replenishment."

"I can see that oceans and seas are being filled by the torrential rains. The island crusts are expanding due to volcanic activity on its surface," Quarkoff observes. "The environment is so tumultuous that huge lightning storms are raging amidst the torrential rains." He muses momentarily. "All of this is a preparation for our co-creative activity in the microcosm of the universe."

"That's right, Quarkoff." Ruah accepts his insight. "All of the ninety-two atom families are laboring together to cocreate the Earth. You are a part of that community. Rejoice in the manner in which you are preparing the Earth for some momentous events." Ruah watches the quark trio as they continue their observations of planet Earth. The ghostly godmother then says, "Let's continue our journey to the fourth rock planet from your Sun."

As Ruah and the quark trio approach this planet, Quarkie remarks, "Why does this planet's surface contain a fine dust with rusty red color?"

"The surface of this planet is rich in members of the iron family. With the help of the members of the oxygen family, they met and mated." Ruah begins his answer. "The result is the compound iron oxide, which is reddish in color." (I recall from my earthly life that this red planet is called Mars for the Roman god of war, which causes the flow of human blood. It is about one-half the size of the Earth and is located about 141 million miles from the Sun.)

"Its crust contains craters, mountains, and valleys, like the first two rock planets. I am puzzled by the high ridges on two sides of some valleys. What caused these to be formed in this way, Ruah?" Quarko asks.

"What do you think happened, Quarko? You're a quintessential quark. What do you say?" Ruah is putting him on the spot.

"Well, I would say that the turquoise ice comets brought water to this planet." Quarko hazards an answer with confidence. "During the early times of this red planet, it was molten. As it cooled and the surface became crusted, water flowed up from the mantle on the planet. Over a long, long period of time, water flowed swiftly in streams through these valleys and wore down the valleys until these very high ridges were formed."

"Bravo, Quarko!" He receives Ruah's compliment proudly. "How did you figure that out?"

"You pointed out to us how waters were brought to the Earth by icy comets," Quarko begins. "I figured this was the source of the water here too. If water were present at one time, it had to flow somewhere, and the valleys were the logical place. I reasoned that they carved out these deep furrows on the surface of the fourth planet from our Sun."

"I would like to name these deep valleys with the high, sheer walls. Let us call them canyons." Ruah finishes his naming. He then turns to Quarkoff and questions him. "What happened to the water on this planet?"

Quarkoff is happy that Ruah has challenged him also. "Well, I notice that this planet has little or no atmosphere. There seems to be a very thin layer of carbon dioxide. This layer is unable to stop the water from being evaporated by the intense radiation from our Sun. So over a period of time, the water vapor in the atmosphere was blown away into intrasolar space." He is feeling pride as he concludes his answer.

Quarkie is quick to make her observation known. "This is not to contradict what you said, Quarkoff. At the top of the planet, which is shielded from our Sun, there is some frozen water and possibly some of the carbon dioxide captured there by the cold surface."

"Let me compliment all of you for your observations and your reasoning." Ruah is overflowing in his compliments to the quark trio. "You are all quintessential quarks, my dear friends." Ruah give names to the top and bottom of the planet. "Let us call the top and bottom of all these spinning planets poles. Now use your imagination. Imagine that there is a long rod extending down through the middle of the round planets from top to bottom. I call this imaginary rod an axis. So each planet spins on this make-believe axis."

Quarko comments not on the naming but on Ruah's compliments. "I could be jealous that you have taken my important title and conferred it on all of us. I will be generous and allow you, Ruah, to do so."

Ruah is taken by surprise by his generosity. "Quarko, you are growing in maturity."

"Where do we go from here, Ruah?" Quarkie asks.

"Let's conclude our journey by visiting the last five gas planets in our solar system." Ruah concludes his remarks.

"On to the gas giants!" Quarko cries out.

Chapter 32
Back to the Past: A Visit to the Five Gaseous Planets

"Look! We are entering a strange and mysterious place in our solar system." Quarkie is surprised and puzzled. "I thought we would see the first gaseous planet. We find all of these large and small gray objects. I do not understand."

The time is October 1, 3.8 bya.

"I have seen these masses before." Quarko is musing. He continues finally. "These are like the asteroids and meteorites that we have seen crashing into our planet Earth at its formation." He pauses again. "These must be asteroids and meteorites."

"All of these masses of asteroids and meteorites are traveling in an orbit around our Sun beyond the orbit of the fourth planet from our Sun," Quarkoff comments.

"Your deductions are correct, my quark friends." Ruah congratulates them. "These are some of the planetesimals leftover from the supernova explosion that caused the formation of your Sun and all of these nine beautiful planets." Ruah continues. "There are different types of planetesimals in this area; let us call it a belt. In the inner belt closest to your Sun are those asteroids and meteorites that are composed of various metals. In their earlier existence, they were melted to more than one thousand degrees, then cooled and melded. The ones in the middle belt have been subjected to less heat and have a different composition. In the outer part of belt are those not subjected to great heat. They contain some of the same materials as the other asteroids and meteorites. They also contain water and some other frozen gases."

"Could some of these large asteroids someday threaten to crash into our Earth and cause some problems?" Quarko is concerned.

"There is always the possibility that one or more of these asteroids could orbit close to planet Earth, be drawn in by the love force of gravity, and crash onto its surface." Ruah answers Quarko's question.

"If one of those very large ones over there (he points to a huge asteroid) impacted our Earth, it could cause devastation for large part of our planet, couldn't it?" Quarko pursues his inquiry.

"Yes, that is always a possibility! It would be a catastrophic event in the story of your Earth," Ruah answers. Then she redirects the conversation. "We will just have to wait to see what happens in future time. Let's continue our journey to the area of the gaseous giants."

The quark trio passes this asteroid belt by moving above its orbit. As they look away from the Sun and toward the outer gas planets, they sight a very large planet, probably the largest of all the nine planets in the solar system. (My previous life experience on Earth reminds me that this planet is called Jupiter, after the chief god of the Romans. It is more than eleven times larger than our Earth, the largest of all the nine planets. It is 483 million miles from the Sun.)

Quarkie tries to describe this huge gas planet. "The exterior of the fifth planet from our Sun is striped. I see lines of beautiful clouds, alternating colors of pink, rust, orange, yellow, tan, and white. As it travels, its orbit it is slightly tilted away from our Sun and is spinning on its axis like the four rock planets but faster than all the rock planets. If I were at one moment standing on one spot on a cloud, I would be facing the Sun for a period of time. Then I would spin into darkness due to being out of sight of the Sun, only to return to light facing the Sun as I continued to spin and return to my original viewing place."

"There are sixteen small planetesimals that are orbiting this planet." Quarko accounts for all of the bodies.

"Let me test you, dear Quarko!" Ruah again challenges him. "What are these bodies called?"

"You told us when we observed the collision of our Moon with our planet Earth that these planetesimals are called moons." Quarko is quick with his answer. "Like our Moon, these moons are held by gravity in different orbits around this large gas planet."

"A quintessential answer, Quarko." Ruah affirms his response. She then says, "My friends, let's explore this giant gas planet."

Ruah conducts them to the edges of the clouds, which are the outer coat of this planet. "The different colors of clouds are due to the composition of the ice crystals, containing different frozen chemical compounds, like methane and ammonia. So the members of the atom families involved in their creation are causing this beautiful color show." The ghostly godmother gives this direction, "Stay close to me. In a moment of time, I am going to transport you into the core of this gas planet."

Suddenly, they arrive at the core. "This core is solid, not liquid. It consists of some dense and lighter members of the atom families." Quarkoff announces his discovery. "This core is like the solid cores of three of the rock planets. Only our Earth's core is hot and molten."

"Let us move up and away from the core." Ruah leads the way. "The hydrogen and helium atom families have created this cold liquid hydrogen, which extends a long, long distance." (I remember from my science class that this ocean of liquid hydrogen extends ten thousand miles or more in depth.)

"This immense ocean of liquid hydrogen that surrounds the small solid core is boiling hot." Quarkoff describes what he observes. As they approached the outer limits of this ocean, he says, "Look, Quarkie and Quarkoff, on the outer surface of this huge sea, this liquid is turbulent, creating mountainous waves. This turbulence is birthing gigantic clouds containing different colored icy chemical compounds, which we observed as we approached this planet." (My science course mentioned that the clouds were hundreds of miles thick.) "These monstrous clouds are being driven fiercely by the rocketing winds surrounding this planet."

"The hydrogen and helium families, along with other atom families, have cocreated a colorful, dynamic planet." Ruah compliments these atom families.

"Why is it, Ruah, that the rock planets lack hydrogen and helium gases? They seem to have collected in these planets at the outer limits of our solar system," Quarko asks.

"Quarko, I think you have the answer. Why do these gases exist so far out in the solar system?" Ruah turns his question back on him.

"Most of the rock planets in the beginning had no atmospheric covering like our Earth now possesses." Quarko begins his explanation. "Therefore, these gases were directed by the solar wind and radiation to the colder, more distant intrasolar and intragalactic space. They were captured by gravity in these regions and collected around these five gaseous outer planets. Being great distances from the Sun, they are extremely cold. The hydrogen and helium atom families, along with a few other, lighter atom families, have created this gigantic, raging liquid hydrogen ocean and other compounds that make up the gas planets."

"You have developed a very logical scenario, Quarko. Congratulations!" Ruah affirms him.

"What would you expect from such an exceptional, queasy quark?" He glows with pride.

"Can we continue our travels through the outer areas of our solar system, Ruah?" Quarkie changes the subject and spotlight off of Quarko.

"Join me in journeying to the sixth planet from your Sun, my dear friends." Ruah's comment is inviting. She and the quark mates move with lightning speed toward this new destiny.

"I can see twenty-one moons orbiting this gorgeous, dusty, brownish yellow planet." Quarkie has been numbering these small planetesimals as they circle this very large gas planet. (My memory from my earthen life recalls that this sixth planet from the Sun is called Saturn, named for the Roman god of farming. It is slightly smaller than the fifth planet from the Sun. It is located about 886 million miles from the Sun.) "I am fascinated by the numberless rings that encircle this planet. They add to the beauty of this gas planet. Of what are they made, Ruah?"

Ruah brings them to the outer rings to examine their contents. "See that they consist of pieces of ice of various sizes, in addition to frozen dust particles and fragmented pieces of asteroids. There are over a thousand of these rings that orbit this planet. While this planet spins at a determined rate, as it moves in its orbit around your Sun, the rings which are bound by the love force of gravity move in harmony with its attracting planet."

"What about the interior of this gas planet, Ruah?" Quarkoff wonders about its composition.

"It is very much like the first gas planet, Quarkoff." Ruah quickly responds. "Like all the gas planets, it has a small, solid core created by the members of the dense metal atom families. This is surrounded again by the gigantic raging ocean of frigid liquid hydrogen, which creates the cloud cover of cold crystals of hydrogen, helium, and other chemical gases. These clouds howl at five times the speed of tornadoes." (I recall that they rage at over 1,100 miles an hour.)

"This planet is slightly tilted toward our Sun, which gives it a jaunty look," Quarko remarks. "This special gaze and its kaleidoscopic rings make this planet my favorite one."

"Yes, quirky Quarko, this planet is full of colorful, cold gas like you are full of gassy words." Quarkoff could not wait to get back at his adversary.

"I will not honor you with a retort for your misrepresentation of my character, you mindless quark!" His biting words sting Quarkoff.

"Stop this right now!" Ruah intrudes herself into the fray. "It's time to move on to the seventh planet from your Sun." (An earthen remembrance tells me that this planet is named after Uranus, the father of Saturn, in Roman mythology. It is located 1.7 billion miles from the Sun and is less than half the size of Saturn and four times larger than the Earth.)

As they travel farther away from the Sun, Quarkie is the first to make an observation. "What an unusual sight. This massive, blue-green gas planet is lying on its side."

"How do you know, Quarkie, that it is laying on its side?" Ruah is testing her abilities of observation.

"I can tell by the rotation of the planet," Quarkie begins her response. "It spins as if it is rolling along the imaginary line of its orbit. Almost all the other rock and gas planets stand upright or are slightly tilted. These planets rotate as if they are standing on their imaginary line of orbit and traveling around the Sun."

"Maybe this gas planet got knocked on its side in a collision," Quarkoff muses. "It does look strange."

"Maybe it just decided to be different and lay on its side," Ruah comments. "The hydrogen, helium, oxygen, nitrogen, and carbon families of atoms have created this emerald masterpiece. Notice that the top pole of its imaginary axis faces the Sun and the bottom pole away from our Sun."

"Could we enter into this planet and observe its characteristics?" Quarko is inquisitive.

"Follow me, my friends." Ruah leads the way. "Notice that the carbon and hydrogen atom families have created this exquisite turquoise outer color. They have created the frozen crystals of methane, which give it its exquisite color. Let us enter into the clouds surrounding this beautiful planet."

As they enter the clouds of methane, ammonia, hydrogen, and helium, the winds are ferocious and cold, four times hurricane force.

"I feel energized by these crazy winds. I am just floating with these capricious breezes." Quarko is enjoying this tempestuous trip.

After traveling a long, long distance through these tumultuous clouds, Ruah says, "As we found on the fifth and sixth planets from our Sun, these clouds are generated by this icy, surging gaseous liquid ocean tens of thousands of miles in depth."

"Does this planet have a hard core?" Quarkie asks.

"Let me lead you with mercurial speed through this liquid sea to the core of this planet." Momentarily, they are at its core. "It is a small, but hard, center made up of some dense and lighter metals." Ruah makes her final remark. "In fact, this seventh planet from your Sun has been created in a similar manner to all the other gas planets."

"Its emerald color, its strange position of laying on its side, and a few dirty rings orbiting this planet are the only real differences from the other gas planets." Quarko summarizes some of this planet's different characteristics.

Quarkie adds to these items. "Like the other gas planets, it possesses moons, seventeen to be exact."

"Since we know the interior makeup of all of the gas planets, let us move forward to the eighth and ninth planets from the Moon." Ruah outlines the last part of their journey. (Again, I recall from my science courses the names of these last two planets. The eighth is named after the Roman god of the sea, Neptune. It is 2.8 billion miles from the Sun, about the same size as Uranus. Pluto was the last planet discovered. Some cosmologists dispute whether it is really a planet. It is named after the god of the dead, Pluto. It is 3.6 billion miles from the Sun, very small, about one-third the size of Mars.)

They are now far past the seventh planet from the Sun and approaching the eighth gas planet.

"The exterior color of this planet is a slightly different shade of blue." Quarkie again is the observer. "I would say it is indigo blue." She pauses, then continues. "Is the blueness again due to its methane gas exterior?"

"Yes, Quarkie. The carbon and hydrogen families that have co-created this methane gas have given it this beautiful color." Ruah again is pointing out the work being accomplished by the other atom families. "The vast ocean surrounding its core consists of very frigid water, methane, and ammonia. Again, being so far from your

Sun, it is extremely cold." Ruah looks toward Quarko. "You would have a wonderful time riding the tenacious winds howling above its ocean. You would be rocketed along with the super tornadolike winds, which are the fastest in all the planets of the solar system." (I recall from my science class that the winds reach 1,500 miles per hour.)

"I counted eight moons orbiting this azure planet, Ruah," Quarkoff says. "Each one is icy cold as it traverses this planet."

"So we are ready to complete our journey of the gas planets." Quarkie realizes that Ruah will now lead them to the outer regions of the solar system. "This last planet is very small and dark. Being so far from our Sun, it is extremely cold."

As they near the planet, Ruah gives a description about how this last planet differs from the other eight. "This is a strange planet. It has a slight atmosphere of methane and nitrogen. Beneath this canopy is frozen methane, carbon monoxide, and nitrogen. Strangely, mixed up with all of these frozen gases are some rock materials. It is extremely cold here."

"As I watch its orbit, Ruah, it has a strange route." Quarko tells what he observes. "This planet is tilted slightly. As it makes its orbit around our Sun, it dips inside of orbit of the eighth planet. This is another crazy planet."

"It has one moon, but it is very large, about one-half the size of itself." Quarkie is amazed. "I would have thought that a moon would be much smaller than the planet around which it orbits." She hesitates and then goes on. "The diversity and individuality of all of the nine planets in our solar system is what constitutes its uniqueness and beauty."

"So we have concluded our journey." Ruah reminds them that this journey into the past is about to conclude. "Take one last look out into the vastness of this expanding universe. All the atom families have been cocreating with us, the Divine Beings. What a wonder of cocreation it is."

"I remember my journey through the vast ocean of intergalactic space. Also our visits to the gigantic galaxies." Quarkie is reminiscing and musing. "Our universe is a magnificent creation. It truly is a work of love. You, Ruah, have been the energizer to all the atom families in their creative work. Your creative energy of love has been a powerful, caring support to all of the ninety-two atom families. Thank you for your support and love for each one of us."

"Yes, our gratitude for your love and care." It is Quarkoff, exuding gratefulness. "I thank you in particular, Ruah, for showing us the immense cocreative activity being carried out within our solar system by the ninety-two atom families. Truly, our universe is energized with a great community spirit of love and cooperation."

"It is time to return to the present," Ruah says. "I am happy that you have come to understand that you are not alone in cocreating this magnificent universe."

Chapter 33
Prokye 2 Grows toward Maturity

"I have a suggestion." Hydrojean is excited. With Nitro and Phoson, she is busy living within the semi-fluid environment of Prokye 2. They have decided to work on improving the latter's ability to move about in its environment. "Do you remember the multitude of comets that we observed while this planet Earth was being formed?" She pauses momentarily and turns directly to talk to her mates.

The time is October 19, 3.0 bya.

"What are you suggesting, Hydrojean?" Nitro is getting impatient with her delay. "Let us hear your idea."

"How did the icy blue comets differ from the gray asteroids?" Hydrojean interrogates them.

"The comets were sleek, with long tails." Phoson draws this remembrance. "The tail seemed to be guiding the comet on its journey. On the other hand, the asteroids were all different shapes, but without tails."

"Could the comets and their tails be the model for Prokye 2 to be able to better move in its environment?" Hydrojean finally spells out her suggestion.

"Hydrojean, the tail of the comet is a good model to begin our work of developing a form of locomotion for our friend, Prokye 2." Nitro accepts her concept.

Phoson is musing about her suggestion. "We were recently able to develop DNA by modifying RNA. Our community effort produced the nucleotide bases used in combinations to construct the DNA template to take over one of the previous functions of RNA." He pauses and then directs a comment to Hydrojean. "A tail! A brilliant idea, dear mate! If we accomplished the above task, I believe we can assist Prokye 2 to achieve the goal of locomotion. Now, let's get started!"

"Before we begin our work, let me make an observation." Hydrojean has observed some interesting visitors present in Prokye 2's semi-aqueous interior. "I notice that there are some free-floating RNA and DNA nucleotide bases drifting nearby. Possibly we can utilize these raw materials in building RNA messengers and constructing new DNA bases which incorporate our plan for a tail assembly. This is a later task to be accomplished after we build and install Prokye 2's new tail assembly."

"This is an excellent idea. These stray nucleotide bases can help us in our work." Nitro affirms her new suggestion. Then he urges them to action. "As Phoson has urged, let us begin our construction efforts to build this new tail assembly."

As they begin their work, Nitro proposes a creative idea to enhance their tail assembly. "We need to build this tail so it is tapered and flexible. Then it can generate a motion to propel Prokye 2."

"This tail will have to protrude from Prokye 2 if it is going to be effective in moving the latter about in its environment." Hydrojean tries to visualize how this new addition will operate. "We will have to attach it to some part of Prokye 2."

"What if we used one of the holes that allow food and waste to enter and exit our bacterial home as the place to attach this locomotive tail when it is completed?" Phoson suggests.

"Good suggestion, my friend," Nitro agrees.

"To move, the tail needs to have a source of power," Hydrojean interjects. "We do not need to build a power plant within her assembly. We can connect this tail to Prokye 2's central power system. Then the tail assembly will be able to move Prokye 2 about."

"Hydrojean, you are really visualizing well the operation of the tail." Phoson is very complimentary. "Let us build this concept into our new creation."

The atom trio, along with the other atom family members, begin the construction of this tail utilizing the molecules, amino acids, and proteins available around them. They build it to be flexible and tapered. They then move the tail assembly through one of the openings in Prokye 2's exterior wall membrane and attach it. Finally, they make the connection of the tail with Prokye 2's power source.

They have now completed and installed the new tail assembly. Using the stray RNA nucleotide bases and assembling other bases from the nearby raw materials, they encode the body plan of the tail assembly onto the four RNA nucleotide bases. Once encoded, the latter becomes the messenger carrying this encoded plan to the home of the Prokye 2's DNA template. The RNA messenger constructs combinations of the four DNA nucleotide bases that encode this new design. Then they attach them to the DNA template.

Nitro informs Prokye 2 about the new tail creation and its function. He asks it, "Prokye 2, please order your power plant to send energy to the tail to make it move." The latter responds. The tail receives its power. "Now signal it to move."

The tail begins to undulate, and its whiplike movement allows Prokye 2 to move around its semi-fluid environment. (I remember from my biology class that this modified prokaryote is called a spirochete bacterium.)

"Prokye 2 is moving under its own power. We have accomplished our mission!" Hydrojean is ecstatic. She had ventured a suggestion to give Prokye 2 the ability of locomotion. Now they have completed their mission. She does not claim personal success. "As a community, we have created this new development, bacterial locomotion. We can be proud as a creative community!"

Ruah had been present to all of this intense activity over these several days of work. She surprises them. "I have watched your intense activity, my atom friends!"

"You were present during all our efforts these past few days, Ruah?" Hydrojean inquires.

"Yes, I observed you accomplishing your mission. Unbeknown to you, I was energizing your efforts. You are the real creators of this new tail assembly," Ruah informs them. "I am merely present to encourage you in your creative efforts."

"No wonder the efforts went so smoothly, Ruah," Phoson affirms.

"What is our next task, Ruah?" Nitro seeks future directions.

"That's for you and your atom community to decide. I will be close, sometimes making myself known, at other times, like this, hidden from your gaze. I will always be there for you," Ruah reassures them.

"That's good to know, dear Ruah." Hydrojean accepts the ghostly godmother's assurance. As the latter departs, full of hope, she says, "Nitro and Phoson, let's move on to the future. We have new missions to accomplish!"

* * *

"Since we have developed this new creation," Nitro says as he begins to explore a new mission, "why do we not try to share this new, undulating tail with other prokaryote bacteria? Maybe we can help them to develop the ability to move around their surroundings."

The time is October 22, 2.9 bya.

"Excellent idea, Nitro." Phoson affirms him.

"This will be our new mission, Nitro." Hydrojean is agreeable. "Since we are agreed, let's pursue it. With Prokye 2's new ability to move around our environment, we can now select a good candidate to approach."

A little later, Prokye 2 points out a very large prokaryote bacterium.

"Thank you, Prokye 2, for your keen observation." Phoson is grateful. "Since this bacterium is larger, we should be able to move into its interior through one of its membrane openings. Then we can tell the atom community there about our creation and ask them if they would like to receive our body plan by transferring our DNA template of this new tail assembly into their DNA code."

"Let's go for it, Prokye 2." Hydrojean urges it to action. "Guide us to an opening in this large prokaryote bacterium. Then enter it. We will explain the reason for our visit to their atom community."

Prokye 2, powered by its undulating tail, approaches its entrance. It then guides them through the opening in the membrane and into its aqueous environment. The atom community within this bacterial being is terror-stricken. Never has such a large creature entered their inner sanctum. They were used to atoms, complex compounds, stray DNA nucleotide bases, amino acids, and proteins entering their domain. The size of Prokye 2 causes them great consternation.

"Have you come to destroy us, oh menacing creature?" they shout.

Hydrojean recognizes their deep-seated fear. She tries to assure them that their entry is not to destroy them but to help them. "We have come to offer you a new creation, an undulating tail that will allow you to move about in your watery environment. You can observe our tail in operation as I speak." She pauses. "Prokye 2, demonstrate how the tail assembly works." Prokye 2 powers the tail. It moves gently in this semi-fluid environment. She adds, "It allows us to travel to areas where we can find food and other construction materials to help us in our daily efforts."

Her reassuring words do not calm them. "Leave our domain immediately, you trespassers. We did not invite you here. Get out. You will destroy us. We cannot feed ourselves and give you nourishment. We will both perish."

"Our undulating tail can help you find all the food you need for yourselves and for us." Phoson tries to reason with them. "We come not to destroy you, but to help you."

Fear clouds their reasoning. "You are a threat to our existence. We order you to leave!" There seems no possibility of convincing them that Prokye 2's atom community means no harm. In fact, they are offering them a very helpful creation.

Hydrojean intuits that this community is adamant in its decision. "We should leave this prokaryote bacterium."

"You're right, Hydrojean." Phoson has come to the same conclusion. "Prokye 2, let us depart this creature."

With the help of its undulating tail, Prokye 2 finds a hole in the creature's wall and moves out into freer waters.

Once they are outside, Nitro comments, "Maybe this is not the time to introduce our new creation to the wider bacterial community. Possibly we need to continue to make some additional modifications to Prokye 2." He seems to have arrived at some new idea. There is an air of confidence as he says, "We need to develop a better system to launch new reproductions of Prokye 2. Maybe we can create a better reproduction system to go along with the new tail."

"This can be our next mission," Hydrojean affirms.

"Let's set out on this mission now," Phoson urges.

Chapter 34
Photie and Prokye 3 Find a New Power Source

"Congratulations, dear Photie!" A familiar voice greets Photie, who is living within Prokye 3. It is Sophia. "You have shown courageous creativity in volunteering to embark on this new mission." She pauses briefly, then continues. "I remember your beautiful prophecy a long, long time ago. You said, 'the future is pregnant with possibilities.' And so it is."

The time is October 24, 2.8 bya.

"Oh majestic mother, thank you, thank you for your congenial compliment," the alliterative master, Photie, responds.

"Greetings, friend Photie!" Elohim is joyous at their meeting. "I remember your beautiful phrase as you described your 5-billion-year-plus journey through our universe. You described what you had observed as 'unimaginable creativity.'"

"So it is." Dabar affirms Elohim's quotation of Photie. "As you remember, fantastic Photie, we, the Divine Beings, have always described the future mission of the universe and its quadrillion upon quadrillion beings as being a journey of creativity. Encouraged by Ruah, your energizer, you are creating the universe."

Photie is proud to be in the Divine Being's spotlight. He is soaking up their adulation. As Dabar speaks of the "future mission" of creativity, a question comes to the mind of Photie. "Oh, dazzling

Divine Beings, what is the next precocious part of our mysterious mission?"

Ruah, the ghostly godmother, responds. "That is for you and your atom mates to decide, dear Photie." After a moment of repose, she continues. "What do you want to do now? You are living in Prokye 3 with many members of other atom families. Could they help you?"

"Oh radiant Ruah, complimentary community can assist in our mysterious mission." Photie feels relief from her concern. "This fantastic Photie, as you call me, tends to be sensitively shy, oh Divine Beings."

"If you can establish a close, friendly relationship with quintessential Quarko, dear Photie, you should be able to relate to the members of the atom families here in Prokye 3," Sophia encourages him.

"Thank you, thank you, sweet Sophia!" Photie grows in courage. "I go forth to forge fresh friendships. My gracious good-bye goes out to you, oh delightful Divine Beings."

"We are always close to you, friend Photie," Elohim reminds him. "You just need to call."

"Remember, Photie, I am with you to energize you in your creative activity with my love energy." Ruah adds her comforting words as Photie moves out to meet some nearby atoms.

* * *

Photie lives within the membrane wall of Prokye 3 in a translucent, semi-aqueous environment crowded with multicolored atoms, small molecules, and complex compounds. New molecule nutrients, free-floating RNA and DNA nucleotide bases, and amino acids are entering through Prokye 3's membrane openings. Waste particles and compounds are discharged through these same holes. Furious procreative activity by way of a cloning technique is going on as the atom families continue to replicate bacterial copies of Prokye 3. Its DNA template home is the source of this continual creativity. The RNA messengers plug into and transcribe a copy of its complete DNA code. They move from DNA's home to the center of its semi-fluid interior. Working from this DNA body plan and using the raw materials in this liquid environment, the atom families construct the replication of Prokye 3's image. Then they release the new being through the membrane to the watery world outside.

The time is still October 24, 2.8 bya.

Photie greets some nearby atoms. "Scintillating stranger, my name is Photie!"

A nitrogen atom is the surprise recipient of this simple greeting. "Good to meet you, Photie. My name is Nitrojoan, a member of the nitrogen family." She had been speaking to a friend when Photie greeted her. She continues. "Let me introduce you to a companion of mine, Oxydon. He is a member of the oxygen family."

"A pleasure to meet you, Photie." Oxydon responds.

"Thank you, thank you, noble Nitrojoan and outstanding Oxydon, for your homey, hospitable greeting." Photie is overjoyed by their friendly reception. He gives them a brief description of his life. "I was birthed as a fancy photon, a yellow package of energy, within the first fantastic second of the birthing forth of our unctuous universe. I traversed the expanding, unfamiliar universe from January 1 to May 10 (first second to 5.4 byich). I then met the Divine Beings, Elohim, Sophia, Dabar, and Ruah."

"You know the Divine Beings, Photie!" Nitrojoan expresses her surprise. "I met them shortly after my birth on February 9 (1.1 byich). I had just been fashioned into a nitrogen family member by the hydrogen and helium families in the core of the nuclear furnace of a red star going supernova. With other family members,

I was given a name and commissioned by Elohim and Sophia to cocreate the universe with them."

"My story is similar, Photie." Oxydon begins his tale. "About May 1 (5 byich), I was also birthed by the hydrogen and helium families in the core of a dying red star. I was blown out into a nearby intergalactic space. I drifted until I arrived and was taken in by a nebula that was located on the outer arm of the Milky Way spiral galaxy. I met Nitrojoan there. On September 10 (10. 4 byich), we were caught up in a horrific supernova explosion in the nebula where we resided. The explosion caused some of the hydrogen and helium 4 gases to begin to spin out a new sun. Ruah's love power of gravity grabbed us and kept us in this solar system. We ultimately fell into a watery area near this seashore."

"Photie, we became part of a new prokaryote bacterium, which we helped to create. Ruah vivified it with her power of love." Nitrojoan adds to Oxydon's story. "With other members of my nitrogen family, Oxydon's oxygen family, the phosphorous, hydrogen, and carbon families, we were able to make a replica of our host bacteria, but only after we had created first RNA, and later DNA. Then we were given a mission to go forth and diversify our cocreations. Subsequently, we were a part of several prokaryote bacteria. From time to time, we would be replaced by other family members and discharged as waste into the watery seashore. We ended up being inside this fluidic interior of Prokye 1, where we were enlisted to become a part of this new being, Prokye 3." Nitrojoan concludes her narrative.

"Dame Fate brought us to this hospitable home, outstanding Oxydon and notable Nitrojoan." Photie fills in part of his story. "Dame Fate also brought me into the vicinity of that same deadly, dying red star. The egregious explosion plummeted me into the midst of the same numbing nebula. Its shocking wave tossed me pell-mell into the path of a neutral, orange hydrogen atom. We collided with such fearsome force that a naive negative electron was dislodged. My august person became a positively charged hydrogen atom. I have been a precocious partner in this same cocreative activity. Yes, Dame Fate has brought us together in phenomenal Prokye 3."

"What is our present mission, my friend Oxydon and my new friend, Photie?" Nitrojoan is inquisitive about the future. "Ruah has urged us to continue to create."

For a time, Oxydon muses over her question. Photie is silent but attentive. Finally, the former says, "I have noticed that we are always concerned about having enough power and energy within the bacterial creatures in which we have lived. What if we worked on building a better system of power, one that is more reliable?"

"Innovative initiative has inspired, omniscient Oxydon." Photie alliterates an exaggeration.

Oxydon is inflated with pride. "Maybe we also need to build a second membrane around Prokye 3. With a more powerful source of power, we require stronger walls."

Nitrojoan agrees. "An excellent idea. We will need to summon the phosphorus family to carry out this task, as they are experts in membrane building."

When this plan is offered to the phosphorus family, they are enthusiastic. They immediately set about to create the double-walled membrane. In the meantime, Nitrojoan, Oxydon, and Photie set about the task of cocreating the elements of a better power system.

Nitrojoan remembers the potency of hydrogen cyanide, a compound of which she had been a part of at one time. She summons three molecules of this compound and urges them to join together to be a part of the new source of power. They are agreeable.

In the meantime, independently, Oxydon and Photie are working on another part of the plan. Photie addresses Oxydon. "Momentous memory reminds me that when we created the first rambunctious RNA, we utilized a sucrosey sugar group and a flashy phosphate group." He pauses and then shyly proposes, "Could we use these giddy genial groups to help power palatial Prokye 3?" Photie manifests a lack confidence in his idea. "Maybe this incredible idea is a dopey dream, my friend."

"Brilliant idea, Photie!" Oxydon gushes forth his compliment. Photie responds with disbelief, surprise, and finally, pride.

"Thank you, thank you, Oxydon." Photie's self-confidence rises. "I am happy you accept my idle idea."

"Let us talk to Nitrojoan and tell her about your idea." Oxydon wants to push the mission along. The former is close by, working with the families of hydrogen cyanide. Oxydon informs her of Photie's idea.

"I have worked with three hydrogen cyanide groups and have created a special compound," Nitrojoan tells them. "Let me join this group with the sugar and phosphate groups that you have formed."

"First, let me join the four-tiered phosphate group with the sugar group." Oxydon obtains the cooperation of this group, which bonds together. "Nitrojoan, now ask your special group to join a bond with the sugar group." They bond together.

This super-sized compound is then tied into the original power system, which sends this new source of power to all areas of Prokye 3. This super-compound brings less power than the original source that has driven Prokye 3 from the time of its creation.

The group is deeply disappointed and frustrated by this failure. Photie speaks up courageously. "Minor modification needs to be made in our daring design." Oxydon and Nitrojoan await his suggestion. There is a long pause while Photie muses. "Imaginative intuition leads me to believe that we should remove one of the four perky phosphate groups. Then it can become a turgid trio more ready to be a better source of potent power."

Oxydon talks to the phosphate community about Photie's idea. They agree and one of the four detaches itself. They inform Nitrojoan.

Nitrojoan has been rethinking the composition of her own base group. She calls in two more members of the hydrogen cyanide family of compounds. They work to reorganize the base group.

"I have modified my base group," Nitrojoan tells Oxydon and Photie. "Let us try again to connect our three groups together." This being accomplished, they connect their creation to the power system. Nitrojoan calls to Prokye 3. "Start the power system and see if our new cocreation produces better power than the original power source."

Prokye 3 sends out a signal to the power system. It transfers power to this new creation. It is immediately obvious that they have been successful. The new compound energizes the different functions within Prokye 3 with much more efficiency and vigorous power. Nitrojoan, Oxydon, and Photie are ecstatic with joy and a profound sense of accomplishment.

"Extraordinary cocreation, my dear atom friends." Ruah is quick to praise their accomplishment.

"So you were nearby, oh ghostly godmother." Nitrojoan is surprised but overjoyed by Ruah's presence. "What shall we call this new creation, Ruah?"

"We will call it adenosine triphosphate, ATP for short, as it is composed of a three-fold phosphate group, a sugar group, and adenine. The latter is one of five nucleotides that usually reside in the DNA templates."

"I am happy you have given us a short name for our new co-creation, ATP." Oxydon responds to Ruah's naming.

"Adventurous ATP!" Photie exclaims.

"Well-named, Photie." Ruah compliments him. His person lights up. "ATP will have an adventurous future as it powers Prokye 3 and the diverse new creations in the future, I feel confident," Ruah concludes.

"Where do we go from here, Ruah?" Nitrojoan says.

"That's for you to decide. You have created this new power source. You have strengthened the walls of Prokye 3." Ruah pauses and questions them. "You have one immediate task to perform or all of this effort will go to waste. What is it you have to do, my dear friends?"

"Oh! The body plan for ATP must be written upon Prokye 3's DNA template." With pride, Nitrojoan answers. She continues. "We must request the RNA messengers to transport this ATP plan and the new design of the double-walled membrane and transcribe them upon Prokye 3's DNA template. So in the future, every new recreation of Prokye 3 and its offspring will know how to build an ATP power system and double-walled membrane. I also believe we should keep a copy of DNA template of this specific ATP creation within the power plant itself."

"Sparkly spoken, nifty Nitrojoan." Photie is swift with his praise.

Nitrojoan, Hydrojoan, and Photie build the four RNA nucleotide bases into different combinations on which the design of ATP and the Prokye 3's double-walled membrane is encoded. Their creation is an RNA messenger, which travels to the home of Prokye 3's DNA template. Using the raw materials nearby, it constructs the DNA combinations of nucleotide bases, encoding the new body plans, and attaches it to the DNA template. Their mission accomplished. They begin a conversation with Ruah.

"Your cocreating journey must continue, dear atom friends." Ruah is urging them forward. "Along with the other atom families, you need to improve Prokye 3. Then, you can decide on other innovations that can help the evolving bacterial community on planet Earth."

The trio affirms Ruah's call Your cocreating journey must continue.

* * *

The time is October 30, 2.5 bya. Prokye 3 is floating peacefully in a watery environment. Suddenly, their peace is transformed into unbelievable panic.

"We are being swallowed up!" Nitrojoan shouts. Fear grips Prokye 3 and all of his resident atom families.

"Oh, dangerous doom awaits us!" Photie cries out. Oxydon seems speechless by the impending danger.

Prokye 3 has just been swallowed by a gigantic prokaryote bacterium. In their experience, the trio has never observed such a large prokaryote bacterium.

"You are in our power." The leader of this prokaryote is menacing. "We plan to destroy you and use you as food and raw material. Your demise is at hand."

"What shall we do?" Oxydon addresses the atom community. "Our destruction is near unless we devise a plan to escape this hellish place."

"Put your tempestuous trust in me, dear friends." Photie is taking the lead. "Proud Prokye 3 and amiable friends, mobilize your soaring strength to get ready to leave this hellish hole when I give you the ominous order." Surprisingly, shy Photie exudes confidence and courage. "I will engage this dangerous demon with weighty words. Do you understand my plan, proud Prokye 3 and atom mates?"

As one, they affirm Photie's plan. Yet, they have deep doubts that any plan will save them from utter destruction.

"Oh powerful one, we come in passive peace," Photie begins. "My name is Photie. Hopeful help we offer you."

"You do not understand, oh idiotic one." The leader belittles Photie. "I have captured you and your mates to annihilate you. Your destruction can give us a new food source."

"Thank you, thank you, oh supreme one, for your compliment. 'Idiotic one' is such a triumphant title." Photie spars gently with this ferocious leader. His mates are starting to understand Photie's plan.

"You, ignoramus one. You don't seem to understand that I am getting very frustrated with you." This devilish leader is losing his patience.

"Thank you, thank you, oh eminent emperor. I don't deserve your wholehearted hospitality and your weighty words of comely compliment." Photie is a picture of patience. "Gracious gratitude for calling me an 'ignoramus one.' No one has ever been so wondrously wonderful as you, elegant eminence."

The leader is becoming angrier with each exchange.

Photie alerts Prokye 3 and his mates to prepare to depart right after the next exchange. "Sudden, stealthy start will be our saving strategy." Photie describes his plan of escape.

"You don't seem to understand my power over you and your mates, stupid one." This ferocious leader continues his verbal assault against simple Photie. "You are a simpleton, an oaf, a fool, a dullard." He is at the height of his frustration. "Send an intelligent spokesperson so I can finally inform you how you will be destroyed."

"Thank you, thank you, oh gracious one. You fill me with princely pride when you call me a simpleton."

Photie pauses momentarily to give the signal for Prokye 3's action. "When I finish repeating his derisive words, desperately dash for the nearest hollow hole in this gigantic giant's membrane."

"An oaf, a fool, a dullard ..."

Suddenly, Prokye 3, with its atom families on board, bolts toward the nearest exist. The escape is accomplished.

Within the large prokaryote, the demonic leader explodes in anger. He now understands that he had really been outsmarted by the "idiotic one."

In the watery safety, a good distance from this menacing, large prokaryote bacterium, the atom community and Prokye 3 are praising the crafty courage of Photie. This simple atom had outwitted a very strong enemy.

"You are our hero, Photie. A brilliant plan, beautifully carried out." Nitrojoan leads the adulation.

Photie does not respond. He humbly but proudly accepts their praise.

Chapter 35
Phosie and Prokye 4 Work to Harness the Energy of the Sun

"I must make an effort to develop some friendships in my new home, Prokye 4." Phosie is musing within herself. "I feel lonely not being with Hydrojean, Quarkie, Quarkoff, and Quarko. Quarko! Yes, I even miss queasy, quirky Quarko and his sarcastic temperament." She smiles at herself as she remembers quintessential Quarko in a positive way. Her musing has brought her to a conclusion. "I am going to introduce myself to the next two atoms I pass in this semi-aqueous world inside Prokye 4."

The time is October 19, 3.0 bya.

Once separated from Prokye 1 through the usual replication process, Prokye 4 drifts gently below the surface of the seashore. It needs to be shielded from the deadly ultraviolet rays emanating from the Sun. With the indispensable assistance of the five atom families, Prokye 4 continues its creative process, producing many copies of itself. The bacterial world is becoming very crowded because of the bacteria's ability to reproduce replicas of itself. I also observe that the bacterial community is beginning to diversify. Prokaryote bacteria of all different sizes (they are all still micro-microscopic in size), colors, and shapes inhabit this watery environment at the seashores.

"Hello, my name is Phosie. I am a member of the phosphorus family." Phosie initiates the contact.

"I am very happy to meet you, Phosie." This atom seems to be very sincere in his greeting. "My name is Hydrojohn. I am a member of the hydrogen family." He turns to a companion. "Phosie, this is my long time friend, Oxyjoy, a member of the oxygen family."

Oxyjoy responds to Hydrojohn's introduction. "I am overjoyed to make your acquaintance, Phosie."

Phosie tells them of her life's journey, while Oxyjoy and Hydrojohn share their life's tale. The "provoking event," the explosion of the red supernova star in the outer arm of the Milky Way galaxy, was the event that brought all them to planet Earth. From their birth, they have known the Divine Beings. They also have all agreed with the latter to help create the universe, and especially home planet Earth, under Ruah's direction. They are mutually surprised to find that the Divine Beings are everywhere.

"What mission were you given, friend Hydrojohn?" Phosie wishes to discover if their mission is more specific than the one she was given.

"It was a very general mission. We were to work to diversify the prokaryote bacteria community." Oxyjoy recalls her instructions. "Ruah reminded us that she was nearby as a source of love power to encourage us in our creative work. The latter told us about the other sources of power that united us, especially the weak power of gravity. We were to use our initiative to develop this diversity and to improve the structure of the prokaryote bacteria family."

"I am very happy to be a creator." Hydrojohn exudes confidence. "I feel very close to our planet Earth, our universe, and all the members of the ninety-two families of atoms."

225

"You are both enthusiastic about our mission. With this spirit, we should be able to accomplish some definite goals when we decide upon them." Phosie is also very positive about the future.

"I believe that we need to increase the size of Prokye 4." Oxyjoy begins to point out some goals. "In addition, I have observed that there is much energy coming from our Sun, which we may be able gather in and utilize to help us accomplish our goal of increasing the size of Prokye 4. At present, we have to hide from the intense rays of our Sun." Oxyjoy has experienced the demise of some prokaryote bacteria that were exposed to the direct ultraviolet rays from the Sun. "In our travels together since the formation of our planet Earth, some groups of prokaryotes have devised a method to live above the level of the water." She then describes this community's method of survival. "Some huge prokaryote communities, I have observed, built mats made up of community members." (I recall from my science course that these bacterial mats are called stromatolites.) "They constructed their mats one level after the other. There were some courageous prokaryotes that volunteered to protect their mated community from the intense rays of our Sun by building a defensive layer. They shielded their community, but at the cost of their own existence. These mats are many, many layers high. This allows them to live above the water, in full view of the Sun, and obtain nutrients."

"The atmosphere above the sea waters is very deadly." Hydrojohn adds to Oxyjoy's account. "Some of our atom friends belonging to some prokaryote bacteria families tried to live on the surface of the water, but these bacteria were annihilated by the intense rays of our Sun. Other atom friends found different bacterial homes. They are assisting other bacteria in diversifying and improving their structures." He pauses briefly and then continues. "To try to turn the Sun's energy into power to help Prokye 4 operate more efficiently and to grow in size are excellent goals, my friends. One more suggestion. Possibly we could help Prokye 4 and the bacterial community learn how to live upon the surface of the waters of the ocean."

"I am intrigued by your observations, ideas, and insights, Oxyjoy and Hydrojohn." Phosie is ready to join them in working to accomplish these goals. "Let's get started on the project, my newly found friends."

* * *

The time is October 24, 2.8 bya.

The trio has been involved in their mission for five days (or 200 million years). They have made progress over this long period. They have not yet discovered how to convert the Sun's photonic energy into power to assist Prokye 4 in its life functions.

"We have been successful in helping Prokye 4 to grow bigger." Phosie points to their progress. "Then we have discovered the need to strengthen its membrane wall that contains the light, brownish, semi-fluid environment within which all the atoms, molecules, amino acids, proteins, and free RNA and DNA bases float and work. So the atom families have labored, especially the phosphorus family, to construct an exterior membrane wall that adds strength to the inner wall of Prokye 4."

"All of these improvements have been transcribed by the RNA messengers upon Prokye 4's DNA template." Hydrojohn continues to outline the improvements. "Let us work especially on the development of a molecular platform to receive sunlight. Learning how to catch the photonic energy, harness its power, and then move this energy inside of Prokye 4 to allow it to work more efficiently is our goal. With more energy, what amazing accomplishments might Prokye 4 produce to move our Earth forward."

"I have an idea." Oxyjoy intuits a possible breakthrough. "Let us develop a molecule centered around one magnesium and four nitrogen atoms. Summon the carbon, hydrogen, and oxygen families to work at assembling a structure around this core." She continues, "Then we will try to capture the photons that travel to Earth from the Sun."

All of the three families try out different combinations. In the end, it is decided to put a special receptor at the end of this molecule. It is made up of one atom of carbon, oxygen, and hydrogen connected together.

Oxyjoy brings forth a distant memory. "I remember Elohim calling this compound aldehyde."

"Let us test our creation." Oxyjoy is anxious to discover if it will work. The special molecule is exposed to the sunlight. Photons from the Sun, bearing red light, bounce off the molecule.

Phosie tries to console Oxyjoy. "Obviously, this molecule and its aldehyde connector didn't work. But do not be discouraged."

"Wait!" Oxyjoy has spotted something important. "Photons carrying violet-blue light have just struck the aldehyde receptor on this molecule. It has energized an amber electron, which is transporting the energy into Prokye 4's power plant." Enthusiastically she shouts, "Success! This molecule works."

"If this molecule is able to capture the energy from the photons carrying violet-blue light, there must be a way of modifying this molecule to receive the red light." Overjoyed by their success, Phosie encourages all of them in their joint cocreative efforts. "I remember how a little modification in RNA was able to help us to cocreate DNA. We substituted a hydrogen atom for a hydrogen/oxygen combination. Maybe we can modify our carbon, hydrogen, and oxygen receptor with another compound."

"I agree with you, Phosie. A slight modification might allow our molecule to process the potent red light carried by the photons." Hydrojohn approves her suggestion. "What modification can we make?"

"I remember that at the beginning of the formation of our planet Earth, some of my atom friends lived in compounds of hydrogen cyanide, formaldehyde, ammonia, and methane." Phosie remembers the early days of planet Earth.

"I am intrigued by the last compound you mentioned." The intuitive Oxyjoy has a suggestion. "Why do we not substitute methyl for the aldehyde? Three hydrogen atoms are connected to a carbon atom to construct methyl."

So they construct a similar molecular platform and create a receptor of methyl. The new creation is readied. Photons carrying the red light strike this new compound. It absorbs the red light and dislodges an electron which, in turn, is energized. The electron carries the photonic energy to Prokye 4's power system. The latter is able to move this energy into a molecular carrier, which transports this energy to different areas and functions of Prokye 4.

"It works! It works!" Oxyjoy and Phosie shout in unison. "We have cocreated a new system to harvest the energy brought from our Sun to help Prokye 4 to operate more efficiently."

"Maybe there is a new function we can develop as a source of food for Prokye 4." Hydrojohn proposes a new mission. "Having learned how to harvest the energy sent to us by our Sun through the photon family, possibly the additional energy can help us to harvest new food sources for Prokye 4."

"The carbon family is famous for having wonderful binding sites that enable it to work with the oxygen and hydrogen." Oxyjoy intuits a plan. "We have compounds of carbon dioxide and water that enter our bacterial home. Let us send some molecule carriers full of energy to break down the two compounds and see what happens."

The carbon, hydrogen, and oxygen atoms in these two compounds are energized by an energy carrier.

"I recognize some of the byproducts of this interaction," Phosie says. "One is sugar, which is a source of energy. This can help Prokye 4 to maintain itself. I observe that there is another byproduct from this reaction."

"What is the other byproduct of this reaction between energy and carbon dioxide and water?" Hydrojohn inquires.

"I see two atoms of oxygen, members of my family." Oxyjoy identifies these new beings. "There are no uses for these atoms within Prokye 4." She muses momentarily; then she continues, "Maybe we should just discharge these oxygen atoms as waste." She pauses. "I hate to think of members of my own atom family as waste. Better to release them to the wider environment where they may serve another useful service."

Oxyjoy does not realize how prophetic her statement will be for the future of the Earth and the continuation of the cocreative process. We shall have to wait to see the result of her decision.

The atom community discharges the oxygen atoms, the byproduct of the above reaction, through one of the membrane openings.

"Now we need to transcribe all of the information about the development of these two molecular platforms with aldehyde and methyl receptors, along with the improvements in the electron transport system on Prokye 4's DNA template." Oxyjoy outlines the next step in their mission. "I also suggest that we keep a copy of the DNA template for our creations in the place where the molecular platforms are located."

The atom trio begins to construct combinations of RNA nucleotide bases that contain the design of the above creations encoded on them. First, these RNA bases, now RNA messengers, build and then encode a copy of the DNA design plan and leave in the place where the molecular platforms reside. Then they dutifully carry these new designs to the home of the Prokye 4's DNA template. Using the raw materials nearby, the RNA messenger uses combinations of the DNA nucleotide bases to construct a copy of the new designs. It attaches this new DNA to the DNA template. Now future replicas of Prokye 4 will carry these improvements.

"We have worked so hard to accomplish this mission." Hydrojohn wants to share their advances with other prokaryote bacteria. "Why do we not visit one of the very large prokaryote bacterium nearby and tell them of our cocreation. It may benefit from our creation."

"Good idea, my dear Hydrojohn." Phosie agrees with his suggestion. "This could help to move our cocreation beyond Prokye 4." She addresses the latter. "Prokye 4, when you come in sight of a much larger prokaryote, enter through one of its holes." Their host agrees.

As I observe Prokye 4 from a short distance, I notice that it does locate a large prokaryote. It enters through its membrane hole. The atom community within Prokye 4 makes contact with the atom community within their host prokaryote bacterium.

Oxyjoy takes the lead. "Dear atom friends, we are here to share some interesting advances we have made." She senses that this neighboring community is friendly and open to hear Oxyjoy's message. "Ruah commissioned us to work at cocreating some improvements within our home, Prokye 4. We decided to try to develop a way of harnessing the power emanating from our Sun that is carried by the photon family."

An oxygen atom from this community recognizes Oxyjoy. "Friend, you belong to the oxygen atom family."

"Yes," Oxyjoy responds. "How do you recognize me?"

"A long time ago, we were part of a prokaryote bacterium working to perfect a way to make a replica of our bacterial host."

The oxygen atom recalls these past events. "I am happy to meet you again." He pauses. "Proceed with your story."

"It's amazing that in the midst of this tremendous creativity all over planet Earth that we should again meet here." Oxyjoy acknowledges her remembrance of being colleagues in the past. She continues her story. "We were successful in cocreating two molecular platforms with special receptors that are able to capture our Sun's energy, which is carried by the photon family. We discovered that we needed an aldehyde receptor that captures the violet-blue light rays and a methyl one that captures the red light rays. The two types of photons that strike these molecular platforms dislodge and energize electrons that are carried to our Prokye 4's power system. The latter processes this power and puts it into a molecular carrier to bring this new energy to where it is needed within Prokye 4 to keep it alive."

"We could use additional energy to power our prokaryote home, especially since we have worked to enlarge it." The oxygen friend seems open to Oxyjoy's offer.

"Probably the easiest way to transfer our new creation would be by sharing our DNA template with you." Oxyjoy offers this suggestion.

"Why do you not send your RNA messenger to transcribe this new function off of your DNA template? Then send this RNA messenger to transcribe your DNA information on our DNA template." The long-lost oxygen friend outlines a plan.

Oxyjoy requests the construction of an RNA messenger to carry out this task. It begins to carry out the mission.

"Stop! Stop!" The Oxyjoy's oxygen friend from the host prokaryote has received a dire warning from the other atom family members. "We are in a crisis. We are running out of food to keep our large prokaryote alive. It seems that you are using our nutrients to help power your prokaryote bacterium."

Prokye 4's community of atoms hears growing cries of desperation from the host community of atoms. The crisis is real.

Oxyjoy immediately understands what is happening. "We understand your plight. Being a large prokaryote, you need much food. When we entered into your semi-aqueous home, we started to draw nutrients from your supply." Oxyjoy realizes that Prokye 4's presence threatens the existence of the host bacteria.

"You will have to leave immediately, my dear friends." Oxyjoy's friend seems apologetic. "We are not very hospitable, are we?"

"We understand, dear friend." Oxyjoy gives Prokye 4 the order to leave immediately. "Good-bye and best wishes."

"Thank you for your understanding and for your quick action, friends. Thank you also for sharing your new creation with us." The host bacterial community is relieved. Hopefully, the danger will be over quickly with Prokye 4's departure.

Prokye 4 acts swiftly. Heading for the nearest membrane hole, it moves through and back into the watery environment outside.

"We left the RNA messenger in our host." Hydrojohn is worried about its fate.

"Since we removed the danger facing our host by departing, maybe our RNA messenger will be able to finish its mission. This would accomplish our goal of transferring our new creation to another prokaryote bacterium." Phosie suggests that even though they had to abort their mission, at least they accomplished their goal of sharing their new creation with others.

"An astute insight, Phosie." Then Oxyjoy shouts, "Mission accomplished!"

Chapter 36
The Quark Trio Embarks on a New Mission

"Our journey to the rock and gas planets was fascinating, Ruah." Quarkie is filled with enthusiasm. "I now realize why we needed to make the journey. We live in such a tiny area of the universe. We needed to see the bigger picture."

Quarkoff inserts himself into the dialogue. "All over the universe, the ninety-two atom families are busy cocreating with you, Ruah, Elohim, Sophia, and Dabar. "What a magnificent creation it is!"

The time is October 19, 3.0 bya.

"Since the first second of creation, our energy of love has driven this creative work." Ruah responds to the grateful comments of the two quark mates. "All of the ninety-two families of atoms have joined in this spirited effort."

"There are some special co-creators among the atom families who have led the cocreative work." Quarko is brimming with self-confidence. Quarkie and Quarkoff begin to seethe with anger. They intuit Quarko's next comment. The quintessential quark continues. "Of course, I am one of those special ones."

"You quaky, egotistical quark." Quarkie is swift and sharp in her response. "We are a community of cocreators." She pauses. She returns to the attack. "Your prideful attitude hurts our unity and peace. When will you ever learn a cooperative spirit?"

Ruah has so often had to be the mediatrix of these verbal battles. Exasperated, she says, "At it again, friend Quarko. Where is that spirit of cooperation you promised a short time ago while on our back-to-the-past journey?"

"Ruah, I did promise cooperation on our mission." He pauses momentarily. "I did not give up the use of my gift of humor and exaggeration."

"Your exaggerations lead to conflict, friend Quarko." Ruah is stern in her demeanor. "So exaggeration is not the same as humor."

"I'll try to reform my behavior. oh ghostly godmother." Quarko seems repentant.

"Greetings, quark mates." Elohim arrives with Sophia. "You had a fabulous trip into the universe and the solar system in particular. Ruah told us."

Sophia takes up the conversation. "Has your trip had any effect on your present attitude about your cocreative mission?"

"Although we toil in a microcosmic world within a prokaryote bacterium all over the universe's space-time, the ninety-two atom families are working feverishly to cocreate too." Quarko is serious for once. "This knowledge encourages me to work much harder in my little sector of the universe. I feel a deep love bond for all other beings on planet Earth and the universe, Sophia."

"Our back-to-the-past journey energizes me to double my efforts to push forward our mission as we live in Prokye 1." Quarkoff proffers his positive comment.

"Speaking of mission, Quarkoff, have you decided what innovations you will undertake within Prokye 1?" Elohim inquires.

"Just recently, we spoke of the need to increase the size of Prokye 1." Quarkoff responds to Elohim's inquiry.

"Why would you increase the size of Prokye 1, friend Quarkoff?" Sophia asks.

"Let me answer that, Quarkoff." It is Quarkie who intrudes herself into the discussion. "I intuit that there will be a need for larger bacterium. I do not have any logical reason to justify such a mission objective. Only my intuition."

"Have you spoken of any other possible modifications to Prokye 1?" Elohim asks.

"I have mentioned to my quark mates the need to build a protective wall inside Prokye 1 around the home of the DNA template," Quarko responds. "The DNA is so very crucial to Prokye 1's future and its offspring. I believe that building a membrane around the DNA's home is a very important mission for us to undertake."

"You have your cocreative work set out for you." Sophia rejoices. "Best wishes as you start to pursue these goals."

The quark mates begin their effort. They seem to be working on the first mission, to enlarge the size of Prokye 1.

"To grow means that there is a need to increase Prokye 1's food intake. This will allow it to build up its power system, which can be used to grow in size." Quarkie ventures her opinion.

"For once you are making sense, lovely Quarkie," Quarko comments positively. "We need to muster additional food to power its growth."

"Thank you for your encouraging, supportive words, dear Quarko." Quarkie is delighted at Quarko's affirmation of her. The mission seems off to a unified, encouraging start.

* * *

From October 19 to October 22, 3.0 bya to 2.9 bya (about 100 million years), the quark trio work diligently to increase Prokye 1's size. The latter doubles its size over that period of time.

"Since we have accomplished our mission of helping Prokye 1 to grow tremendously, it might be well for us to work on our second cocreative goal, to build a protective wall for its DNA template home," Quarkie advises.

"We need the help of the phosphorus family of atoms to accomplish this mission," Quarkoff responds.

"I have anticipated your planning, my esteemed quark mates." Quarko is ready. "I contacted our phosphorus families and they are ready to cooperate. They have a lot of experience at membrane building."

"Good work, Quarko." Quarkie is complimentary of his foresight.

The phosphorus family begins to construct the wall around the DNA template containing the encoded body plan of Prokye 1. Within a relatively short time, they have completed their task. Ruah is present for the last part of the construction.

"Ruah, is there any name that we can assign to this membrane wall within which the DNA template resides?" Quarko asks.

"Since it is located inside of Prokye 1, it is similar to the location of the protons and neutrons in your atomic structure. Let's call this new being the nucleus." Ruah gives it a name.

"Excellent name, ghostly godmother. Just like we quarks who make up the protons and neutrons in an atom, our place is crucial to the functioning of an atom," Quarko concludes. "So, too, the new home of DNA is very important to the life of the prokaryote bacterium."

"Since we have constructed this protective wall around the DNA template, I believe that we need to reorganize its thousands of nucleotide bases into a better format." Quarkoff is the initiator of this novel idea. "I foresee that we will be adding to the number of the DNA nucleotide bases as time goes on."

"What kind of a structure would you use to organize these bases?" Ruah asks.

"I have been giving a lot of thought to this since there are so many bases in the DNA template and more will be added later." Quarkoff continues, "I know we need a structure that will be efficient and effective." He pauses and then says, "I believe the nucleotide bases should be housed and organized into a spiral helix."

"Brilliant idea and plan, Quarkoff!" Quarko shouts. "This format will serve prokaryote bacteria as they grow into the future. It will allow for later expansion."

The quark mates begin the tedious work of reorganizing the four DNA nucleotide bases into a spiral helix. Quarkoff comments, "I also believe this spiral format will more easily assist the RNA messenger to plug into the DNA template and do the encoding work."

"We have accomplished not a dual mission, but thanks to Quarkoff, a three-fold mission." Quarkie is exultant. "Ruah, can we now rest from our labors?"

"That is for you to decide, dear cocreators." Ruah is emphatic.

"Now that we have doubled Prokye 1's size, I think we should invite some smaller prokaryotes to make a home with us." Quarkoff makes still another bold proposal. "Possibly they might be able to utilize that additional interior space and add some new functions to Prokye 1."

"Not so fast, Quarkoff," Quarkie cautions. "We still need to encode these new body plans for doubling Prokye 1's size, for constructing the nucleus membrane and the new spiral helix design upon an RNA messenger. Then we need to send it through an opening in the DNA nucleus to add these plans to the DNA template."

"Important suggestion." Quarkoff accepts her immediate plan of action.

The quark trio constructs an RNA messenger from the free-floating RNA nucleotide bases and other bases made up of molecules and atoms in Prokye 1's semi-fluid interior. The RNA messenger journeys to and then enters this new nucleus through an opening to add these new design plans to the DNA template. Quickly, this task is accomplished.

"Now that we have preserved these new adaptations on the DNA helix, I again make my proposal. Let's invite a smaller prokaryote into our large space and find out if they would be able to offer Prokye 1 some innovations that would improve its functioning." Quarkoff is insistent.

"I do not like this idea." Quarko dismisses his suggestion. "We might end up harming ourselves or even endangering Prokye 1's existence by such a bold move."

Surprisingly, Quarkie agrees with Quarko. "I also think that your move is too bold and carries too many risks. I am strongly opposed to such a dangerous move. Let's play it safe."

"Since there are two of you in opposition, I will have to withdraw my suggestion." Quarkoff decides not to fight for his novel idea. "But I still believe strongly that if we are to be co-creators with the Divine Beings, we need to act boldly at times to push creation forward."

Ruah has been overarching this conversation and controversy. "I commend you, Quarkoff, for your forward thinking." Ruah pauses. The quark mates eagerly await her conclusion. "I compliment you for being able to agree to disagree and to search for consensus on this issue. This type of behavior is what cocreation is all about."

"Let's move forward into the uncertain future of cocreating with deep hope and unity of purpose." Quarkie ends this event with a note of optimism.

Chapter 37
Back to the Past: A Journey through the Universe

"Greetings, my dear Nitro, Hydrojean, and Phoson!" Sophia exudes joy. "Elohim and I are here to take you on a journey back to the past."

Hydrojean is taken by surprise by the title of the journey. "What is a journey back to the past?" She appears confused.

The time is October 22, 2.9 bya.

"Do you remember, good friends, how we defined time at the beginning of the universe?" Elohim questions them.

"In time, there is a past, a present, and a future," Hydrojean recalls.

Phoson continues. "We are now living in present time, October 22, 2.9 bya." He pauses and ponders. All wait for his next words. "So … so you, Sophia and Elohim, want to take us back to past time."

"Excellent reasoning, Phoson." Sophia commends him.

"Why go back to the past?" Nitro inquires. "We are presently involved in trying to perfect Prokye 2's tail assembly so that we can help it move through its aqueous environment more efficiently."

"In addition, we are very busy trying to create a more effective way to launch the new mirror images of Prokye 2 into the watery environment outside of it," Hydrojean interjects. "We are close to a breakthrough in present time."

"My dear atom friends, let me explain the reason for this journey," Sophia begins. "You live in a tiny, microscopic community, a microcosm. Elohim and I would like to introduce you to the larger community of the universe, the macrocosm, from which you descended."

"Your multitudinous friends in the ninety-two atom families have also been very hard at work in cocreating." Elohim adds to Sophia's rationale for the journey. "As you labor in your microcosmic world, you are unaware of all of the marvelous cocreation going on in the macrocosm of the universe." He pauses to allow their reasoning to be absorbed.

The latter are in deep thought. Finally, Phoson responds. "Elohim and Sophia, this sounds exciting. It is true. We are so immersed in our present project that we are cut off from the immense amounts of creative activity going on in this gigantic universe."

"You are right, Phoson," Nitro agrees. "I am excited too. Let's get ready to go!"

"Wonderful!" Hydrojean is enthusiastic. "Let's go back to the past!"

"Follow us, my dear friends." Sophia gives an invitation to begin the strange journey.

In present time, it is October 24, 2.8 bya.

* * *

"I want to first take you back to September 10, 4.6 bya," Elohim explains. "I want you to be present to the origins of planet Earth."

"Sophia, protect us!" Hydrojean shouts. An immense red star has just gone supernova. The radiation and stupendous amounts of debris shoot out into space. Suddenly, Hydrojean remembers. "The 'provoking event' you intuited some time before this event, Sophia."

"I can see our Sun being spun out within the nebula nursery by the hydrogen atom and helium nuclei families." Nitro is excited to observe this tumultuous event being lived out again. "I see the nine planets beginning to form by the accretion of asteroids, comets, cosmic dust, and other debris from the horrific explosion."

"It is hard to believe that the third red fiery accreting planet from our Sun will someday become planet Earth, on which we live and cocreate." Phoson expresses his amazement at the very early sight of the reddish, molten planet Earth.

"First, I want to take you on a speedy tour of your Milky Way galaxy." Elohim points out the direction of their journey. "Follow me. I want to show you this beautiful galaxy from above." The Divine Beings lead them to a vantage point well above the central core of the galaxy.

"What a magnificent sight." Hydrojean is enthralled. "I can see the black hole in the middle of its core." She stops for a moment. Grief begins to fill her being. "I am reminded of my strong-willed friend, Hydro, when I see this black hole. From the very early days of our universe, I recall with horror his pride-filled, obstinate nature. Hydro was warned personally by Dabar to stay away from the threshold of black holes."

"What did he do?" Phoson is interested.

"Hydro was upset that the Divine Beings did not have an overall plan for the creation of the universe." Hydrojean begins her response. "He believed that he could take over leadership in the work of creation. One day, when he and I were traveling through the universe within a galaxy, Hydro sighted a black hole that seemed to be consuming everything near its threshold: space dust, gases, stars, and anything that came nearby. Its drawing power was stupendous. Hydro was curious. He moved in the direction of its all-consuming, black opening. Before I realized, I was also being sucked toward this hole. Fortunately, I struck some object, which hurdled me away from this imminent danger. Hydro continued to move toward the threshold rim of the black hole. Too late did he realize that his foolhardy venture would end in disaster, Phoson. I heard him cry, 'How stupid I am! I am finished!' I saw him drawn down into the vortex of its monstrous, cyclic winds."

Elohim takes up the story. "Once sucked into this tornado-like cauldron, Hydro was torn apart and virtually annihilated." He pauses. "Poor Hydro. He refused to be a part of the co-creative community. His egoism and pride led to his demise. The community of quarks and atoms were deeply grieved at the loss of their foolish friend."

"We have survived our grief, Phoson, and have worked with the encouragement of Elohim and Sophia, along with Ruah and Dabar, to cocreate the universe and planet Earth." Hydrojean seems to experience a catharsis from telling her sad story. "I now discover, Phoson, that you are involved in the same mission."

"In the canopy over the galaxy's core, there are multimillions of yellow suns. Most of the other suns are blue-white in color. A few of the others are different shades of red. What is the significance of the color of a sun?" Nitro inquires.

"The yellow color of a sun is a sure sign that it is older." Elohim answers his inquiry. "The blue-white color indicates that these are younger suns. What does the red color indicate?"

"They are dying suns. The more monstrous ones will end up as supernovas." Nitro remembers the one that sparked the creation of the solar system.

"Other red suns will end up as flashing neutron suns, red dwarfs, brown dwarfs, or even black holes." Sophia adds her comment.

"How beautiful are the Milky Way's five majestic arms, filled with billions of nursery nebulae that create the billions upon billions of suns that are replacing the aging and dying suns. The nebulae are kaleidoscopic, with a blending and overlapping of magnificent reds, blues, greens, yellows, and browns." Hydrojean is filled with awe at this scene. "The core, and its canopy, the five flowing arms extending from the core, seem to be laid out on a rotating disk. The love power of gravity binds these hundreds of billions of suns into a galactic community."

"I observe that this galactic community is moving out into expanding space, and each star is moving gently away from each other as it journeys into the future." Phoson exults at the sight. "Every sun is rotating on its imaginary axis, and the whole galactic community is rotating rhythmically around the gyrating, menacing black hole in the middle of the core." Nitro is awestruck at the sight. "Due to the special vision with which you have gifted us, I can see our solar system, our Sun, and our home, planet Earth."

"Let's leave your home galaxy and travel at unimaginable speed to different places in this beautiful universe." Again, Sophia leads the trio at breathtaking speed, way beyond the speed of light.

"Wow! This is amazing." Hydrojean is enthralled by the speed and the view. "The blackness of intergalactic space is amazing. In the middle of this ebony space, each of the hundreds of billions of galaxies appears like lighted communities."

"We are approaching the far edge of the ballooning universe." Elohim reaches its outer limits. "Dear friends, it is September 10, 10.4 billion years into creation history. If we were to travel from here back to the first moment of January 1 and the birthing in love of the stupendous universe, we would have to travel 10.4 billion years. Our photon friend, Photie, presently journeying with Prokye 3, could travel this distance if he moved at his photonic speed of 186,000 miles per second. I would calculate the distance Photie would have traveled to be 10.4 billion light years."

"Here at the edge of the universe, I hear a strange, soft, melodious echo." Phoson is in awe of this sound. He turns to Elohim and says, "From where is this echo emanating?"

"When we sent out Ruah to birth the universe, there was a magnificent, stupendous explosion of our love energy." Elohim begins his response. "This rapturous energy was so intense that it created an echo that will continue to resound as the universe expands." He pauses. "Phoson, you are in contact with the very beginnings of the universe's journey of love." (I recall that this is the cosmic background radiation first discovered in 1965 by Bell Laboratory scientists.)

"Awesome, Elohim!" Phoson is fascinated. "In touch with our universe's birth. Wow!"

Hydrojean is looking in another direction. "I notice some strange stellar masses, Sophia." She comments on a nearby site on the edges of the universe. "Nearby is a blindingly brilliant body emitting many types of radiation: red, violet-blue, and yellow. What kind of a body is this?"

"About 8 billion years ago, two galaxies crashed violently into each other, creating an immense black hole." Elohim recounts its history. "This black hole began to consume the stars, dust, and debris around it. The temperature was so overwhelming that this body emitted enormous quantities of brilliant, invisible radiation, but little or no light. When you view it from a side angle, the love force of gravity is so enormous that very little light from the stars and debris entering the black hole can escape. Thus, it receives its name from the blackness of the space around its entrance."

"What will you name this body, Elohim?" Phoson is fascinated by its brilliance and vitality.

"I name it a quasar, since it gives off a special signal we call a radio wave." Elohim finishes his naming and adds a comment. "If we were back at your solar system of the Milky Way, this quasar's light and radiation would have to travel about 6 billion light years to reach you. You would observe it as a very dim light. That tells you something, Phoson, about the immensity of our magnificent universe."

"Let's start back toward the Milky Way." Sophia again leads the trio.

Passing a nearby galaxy, Nitro spots a body that is spinning with tremendous velocity and emitting intense radiation of various kinds from opposite poles. He pauses and directs a question to Sophia. "What kind of a body is this? How did it end up in this way?"

"Originally, this was a medium-sized sun." Sophia begins her historical account. "When it used up all of its helium nuclei fuel, its masses of gas and other atoms crashed down on its core. It crushes the protons and electrons together to form neutrons. It emits very lightweight particles of neutrinos into space. In the meantime, it develops magnetic fields that emit the light, spinning so rapidly the light appears to be blinking on and off."

"Because it consists of a mass of neutrons, I will call this a neutron star or sun." Elohim names the body.

"It was its blinking lights that caught my sight." Nitro is still amazed by its light action.

"If you look in the opposite direction in this galaxy, you will find the remains of another sun." Elohim takes up the discussion. "It is a large black hole. At one time, it was a tremendously large star. It went supernova, blowing off all of its gases and other elements. The pressure exerted by the implosion created this black hole that you observe. So dying suns can and do end up in different forms: neutron suns, black holes, white and brown dwarfs."

Quickly, at super-speeds, Elohim, Sophia, and the trio traverse the almost infinite distance from the edges of the universe to the Milky Way galaxy and just above the third planet from the Sun.

"Now, let us take a closer look at your Earth home." Sophia comments upon their arrival at the Earth. "Just as you observed the activities of the ninety-two atom families in the wider universe, let us observe what they are doing in and around your home planet, under the surface crust, on the crust, in the oceans, and up the atmosphere."

"Lead us on, dear Sophia and Elohim!" Hydrojean shouts with exuberance.

Chapter 38
Back to the Past: A Journey into and around Planet Earth

"Our Sun continues to grow in intensity." Nitro observes that it seems brighter and stronger than when he saw its nuclear furnace ignite on September 10, 4.6 bya. "At present, its intense radiation is not able to fully penetrate to the surface of the Earth."

The time is September 16, 4.4 bya.

"The members of the carbon and oxygen families have created a cloud cover of carbon dioxide to shield the Earth from our Sun's intense radiation." Phoson reports what he sees. "I notice that our Earth's surface is a fiery conflagration due to the fueling activities of the volatile uranium and thorium atom families. Look at those gray asteroids of various sizes, azure-colored comets with long, icy tails, and meteorites slam into the Earth. They are bringing many of the atom families that will be needed to build the Earth."

"The atom families have created the red-hot magma beneath the surface that is belching into the atmosphere huge geysers of this fiery liquid silica, water ferried to our planet by the icy comets and various gases." Hydrojean adds to her observations. "This is what is causing the atmosphere to be so cloudy and overcast, blocking much of our Sun's radiation. The waters are collecting as vapor in the air. At present, I believe the air is too hot for the water to condense and rain down upon the surface of the Earth."

Sophia comments, "Notice that there is little water upon the surface of the Earth at this time. In most places, the surface is fiery and molten due to the continuous bombardment from outer space."

"Let's move forward from September 18, 4.3 bya." Elohim escorts them. "Do you notice any changes on the Earth?"

Hydrojean is the first to respond. "I notice that the cloud cover has thinned. The air is extremely warm, and the few small basins of water that exist on the Earth's surface are still very hot." She looks around to continue her observations. "The bombardment from asteroids, meteors, and comets is slowing down. Around some belching volcanoes, more of the surface of the Earth is being paved. These crusted areas exist like islands amidst the still volatile, liquid lava."

"Let's enter into the raging hot magma and visit with the atom families who are creating this important material that will help the Earth to progress into the future." Sophia, with Elohim, leads their atom friends into the bowels of the Earth. They are able to penetrate the boiling magma and are not scalded by its intense heat. "Notice that the islands of solid crust are floating upon the surface of this seething, restless underground magma."

"Greeting to you, dear atom friends." Sophia greets some of the silicon and aluminum atoms. "How is your work proceeding?"

"We are happy to see you once again, dear Sophia and Elohim." One of the silicon atoms greets the Divine Beings. She goes on to answer her question. "We have labored to use our lighter atom families to fashion this ruby liquid, which we believe will be important in creating and maintaining the crusted surface and the rocklike floors of the water basins. The uranium, thorium, and other volatile atom families have contributed their fiery energy to this task."

Hydrojean is curious. "So you know the Divine Beings too?"

"We have known them from the time of our birth into existence and have been empowered create the universe and Earth." She acknowledges their long acquaintance.

"I get the impression, dear silicon friend, that there is a constant exchange of atoms, along with solid and gaseous molecules, through the vents or cracks on the surface of the Earth. How does this help the creation process?" Nitro asks.

The silicon atom explains its importance. "If we merely produced our magma and sent it on a one-way trip to the surface of the Earth and atmosphere, we would soon run out of atom members." There is a slight pause. She continues, "We need to recycle the materials after they have done their job on the surface of the Earth. So through the same cracks we use to send out our products, the atom families that make up water, gaseous compounds, and other materials come back to this area to be recycled and reformed again. It takes a lot of teamwork to make this process work. We have splendid cooperation from all of the atom families who live here."

"My friend, we of the bacterial community of atoms work in a like fashion." Phoson sees a similarity to this recycling process. "We use many of the compounds constructed by the atom families in our environment to help our host, Prokye 2, to live and grow. We do generate waste, which we return to the watery environment outside of Prokye 1 so other atom families can reuse this material, recycling it."

"What is the bacterial community?" The silicon atom is very curious.

Phoson immediately realizes that they have been traveling back to the past. He decides not to try to explain this to his new friend. "It's too long to explain. We can talk about this some other time, my friend." The latter does not pursue an explanation.

"It's time to move forward again, my dear friends. We wish to thank you in the silicon and related families for your hospitality to us." Sophia again acts as leader. She directs her next comment only to the atom trio. "Let's move forward in the past to September 23, 4.0 bya."

Having arrived at this point in time, Nitro comments, "The crashes of asteroids, meteors, and comets are very few and far between. The size of these bodies now is much smaller so the disruption of Earth's surface is minimal. With the help of the love force of gravity, our planet has almost swept clean all of the space debris near it."

"This is allowing the surface to harden in many places." Hydrojean is checking out the volcanic activity. "The volcanoes are everywhere on the surface of the round Earth. I notice that the depressed areas of the Earth's surface have been paved with a blacker type of material."

"Sophia, the atmosphere around our planet is beginning to cool off." Phoson is surprised at what he now observes. "In this cooler atmosphere, water vapor is condensing and beginning to fall upon the planet's surface and is collecting in those depressed paved areas amidst active volcanoes."

"What is that loud noise that we hear in the air, Elohim?" Hydrojean stops to listen. "There are narrow, yet gigantic streaks of light that accompany the ponderous sound."

"Let's travel up into the clouds and ask the atom families why this light and noise are happening," Elohim says. Then he directs the trio upward into the atmosphere.

There, they are greeted by some members of the hydrogen and oxygen families, who created water, and the carbon and oxygen families that created gaseous carbon dioxide. Elohim initiates the conversation. "My atom friends, I would like to introduce you to Phoson, Nitro, and Hydrojean."

"How good to visit with you again, Elohim and Sophia." One of the hydrogen atoms initiates the greeting. The traveling trio is surprised to again learn that these atoms are personally acquainted with the Divine Beings.

"My trio of fellow travelers has observed the large streaks of light and has heard some violent noise." Elohim says. Then he poses a question to these resident atoms. "What is the source of this light and noise?"

The oxygen atom then begins to explain their causes. "My friends, you are aware, are you not, that our planet Earth is rotating at a constant, rapid speed upon a mythical axis as it orbits our Sun. This swift movement causes strong winds to be generated. The atoms of hot gases rise up and are battered in the fast-streaming winds. They tear some of our electron friends away from orbit around the nucleus of our atoms where our quark mates live and work in our neutrons and protons. The freed electrons have a negative charge, and the nuclei of atoms, devoid of an electron, are positively charged. Inside of these furious wind storms, the negatively charged electrons particles come in close contact with us, the positively charged atoms—remember, we are talking about quadrillions of negatively and positively charged particles and atoms—and generate these long, power-laden streaks of light and this loud, ponderous sound."

"Elohim, what name do you give to this light and sound?" Nitro poses the question.

"I name the gigantic streaks of light, lightning, and the deep ponderous sound, thunder." Elohim responds quickly.

"I have another question." Hydrojean is intense. "Some days ago, I noticed that the water vapor stayed in the atmosphere. Now it is beginning to fall to the surface to fill the depressed basins. What has changed, my dear atom friends?"

One of the carbon atoms takes up this challenge. "About two weeks ago, our carbon dioxide family was very numerous in the atmosphere. These molecules were generated by the fires raging all over the surface of our planet. As the fires have diminished due to less space debris striking its surface, the Earth has begun to cool. Heat tends to rise. So as the coolness of the air below meets the hot gaseous air that is full of water vapor, the latter condenses and drops as rain. These waters begin to fill the depressed basins on the planet's surface. The more the Earth cools, the more this process operates. The rains, carrying our carbon dioxide, descend in a steady stream to fill the basins of the planet. The more persistent and violent the clashes between the upper winds and the atom families which live there in gaseous and solid form, the more charged particles there will be. Thus, you will have larger and larger lightning strikes and louder and louder thunder claps." He pauses. "Later in your journey, you will see how this will come true."

"We look forward, my friend, for the fulfillment of your prophesy." Hydrojean is truly grateful. "Thank you for your explanation of this process."

"Thank you too, dear atom friends. See you soon." Sophia is congratulatory. She then turns to the trio and says, "Off we go to the next time period on our journey, September 30, 3.8 bya."

As the Divine Beings and the atom trio move forward at fantastic speed, they are able to observe the minute changes going on day by day. Finally, they are present to planet Earth on September 30.

"Our atmospheric atom friends were right." Hydrojean has observed the carbon atom's prophecy come true. "All the way along our journeying, I could see the rains growing in intensity. The lightning bolts are more massive in size and power." She pauses and assesses their present environment. "Wow, the rains are falling in torrents all over the surface of the Earth. How awesome and gigantic the lightning bolts flashing across the sky! The stupendous thunder claps. A magnificent experience!"

"Since we are back to the past, we are part of the first living beings to see the energetic lightning and to hear the roaring thunder," Nitro muses. "When the first lightning bolt struck and the thunder rolled over the atmosphere, there were no living beings existing. Amazing is our journey into the past."

"Let us enter one of these bodies of water." Elohim is showing the trio the way. "By the way, I have taken the quark trio back to the past and named the bodies of water. The very large bodies I named oceans, the medium-sized ones, seas, and the small-sized ones, ponds or lakes."

Elohim leads them to the tempestuous edges of the agitated ocean. "Let us discuss the makeup of the waters with some of the hydrogen and oxygen atoms that live here."

"Could we talk with you, dear hydrogen and oxygen atoms?" Elohim initiates the dialogue.

"What do you know, dear Elohim and Sophia, Nitro, Hydrojean, and Phoson, we meet again." It is the hydrogen atom that takes up the conversation. "Dear atom friends, we met you near our atmospheric home on September 23."

"It is our friends from the near past time!" Hydrojean shouts with jubilation. "It's good to again meet a member of our hydrogen family working to cocreate the Earth. I feel deep pride in our hydrogen family."

Sophia takes up the conversation. "I think that our atom trio has some question to ask about the waters of this ocean."

"Yes, Sophia, I want to know what atom families are dwelling in these watery oceans, seas, and ponds." Phoson begins the questioning.

"You are aware, are you not, that there are many types of gaseous molecules in the air." The hydrogen atom is testing their knowledge.

"Yes, we do." Hydrojean recalls her experience. "I was once a member of a formaldehyde molecule family. Nitro was a member of the ammonia family. So we know of the many varieties of molecules in the air."

"We were part of the water vapor that condensed into rain," the hydrogen atom recounts. "As we fell, we picked up as passengers many kinds of molecules. As we landed in this ocean, we deposited our molecule friends in these waters. So this seashore is rich with many and varied kinds of molecules." After a few moments' delay, she adds, "Our water molecule friends that have fallen on the hard, earthen crusts found pathways so that they could return to the oceans and seas. As they traveled, they snatched from the crusted surface valuable atoms and molecules, including sodium, and carried them to the open bodies of water, further enriching them."

"Sophia, what name do you give to the pathways that carry the rain waters over the earthen crust to the oceans and seas?" Hydrojean asks.

"I name these pathways, rivers," Sophia responds quickly. She adds, "They carry in their waters many important atoms and molecules to the oceans and seas. This helps the atom families in recycling themselves in the creation of the planet."

Phoson is anxious. He wants to cry out and let these new friends know that it was the rains that brought him and his friends to the shores of an ocean like this. He decides to hold back this information since it happened in the future on October 4, 3.6 bya. How could he possibly explain to them a journey into past time?

With enthusiasm, the oxygen friend blurts out, "All of us living here in these waters feel that some momentous event is about to happen. All of this accumulation of varied molecules created by the atom families must be preparing for this event."

"An interesting prophecy, dear friend." Sophia compliments her insight. She also does not want to confuse these two atom friends by telling them that their prophecy did come true. "Well, thank you for giving us some information about the composition of these powerful waters and their inhabitants." The Divine Beings and the atom trio bid them a farewell.

As they ascend to a vantage point above the Earth, they stop. Elohim asks, "Do you observe anything new that has occurred since we last viewed planet Earth from here?"

"More of the radiation from our Sun is able to enter through the atmosphere and strike the surface of the Earth," Phoson observes.

"Why would that happen, Phoson?" Elohim asks.

"The torrential rains have washed many of the carbon dioxide molecules out of the air and deposited them in the oceans," Phoson reasons. "This has allowed more radiation to enter onto the Earth."

"Excellent reasoning, dear Phoson." Sophia gives a compliment. Phoson is filled with pride. "What other new occurrences have happened to the planet recently?"

"More of the surface of the Earth has been paved with cooled silicon magma." Hydrojean looks over the hardened crusts and the floors of the oceans and seas. "Probably the hardened crusts constitute about 10 percent of the planet surface at present, while the oceans, seas, and ponds represent 90 percent. I believe that the hardened crusts are growing in size and proportion as time goes on."

"An astute observation, dear Hydrojean." Sophia compliments her also. There is a period of silence as they look over the planet. Finally, Sophia says, "Let's journey forward to October 4, 3.6 bya, a momentous time in your existence, dear atom friends."

Chapter 39
Back to the Past: A Return to Prokye 1's Birthplace

"I recognize this place!" With joy and exultation Hydrojean shouts. "This is the birthplace of Prokye 1. I recognize the clay material. There is the bubble home created by Phosie and his phosphorus family. I see Phosie now." She continues her perusal of this site. "There you are, Nitro."

"I see you, Hydrojean." Nitro is overjoyed. "Quintessential Quarko and his two quark mates are also present."

The atmosphere surrounding this scene is calamitous. Elongated ribbons of light flash as the driving upper winds agitate the particles, displacing electrons from their orbit around the atom's nucleus. The negatively charged electron particles clash vehemently with the positively charged protons and neutrons to produce this energetic light show. The bolts of lightning strike close to the slimy, molecular rich seashores where the atom families are working feverishly within the clay medium within their bubble-wombs. The winds and rains agitate the waters of the sea. These waters are hot and acidic.

"Sophia, can I talk to Phosie and myself?" Hydrojean wants to enter into a dialogue with herself and friends in the scene she is observing.

Sophia hesitates to find out if Hydrojean will answer her own question. The three atom mates, Hydrojean, Nitro, and Phoson are silent. The pause is broken by Phoson.

"That's impossible, Hydrojean." Phoson is deliberate in his speech. "We have gone back to the past, so there can be no dialogue with ourselves. On this journey, we have dialogued with some of our atom families to ask them to explain a creative process. With ourselves, we have to refrain."

"We were really laboring long and energetically in putting the molecular compounds together in the watery chemical soup at the edges of the ocean." Nitro is intrigued at watching himself and his atom mates. "Look, Ruah is overarching this bubble and us, its inhabitants. Her loving touch is very intimate to this nascent bubble." His excitement grows as he carefully observes this bubble in its clay womb. "It is vivified and enlivened by Ruah and comes to life."

"You too, my friends, were crucial to the enlivening of this bubble." Elohim reminds them that on this crucial occasion in the history of the universe and planet Earth, they are cocreators with Ruah. "On the physical and chemical level, you assembled this complex community of molecules, and the atmosphere contributed the energetic lightning. Ruah vivified the bubble at a deep level. How wonderful that you have been true to your word and have cooperated in the process of the creation of the first living being." Elohim pauses and turns to Phoson. "You were present nearby in a community with other atom families, cocreating with Ruah at the same time that Prokye 1 was enlivened."

"It was a privilege to cocreate with Ruah, my dear Elohim and Sophia." He appears justifiably proud.

"Let's travel to the ocean depth." Sophia and Elohim conduct them speedily to a vent in the ocean floor. Sophia says, "Here, the atom families are involved in the same cocreative task: to produce biological life."

"The temperature here is very, very hot." Phoson notes the environmental conditions. "The volcanoes are spewing out a very dark and dense magma."

"I will name this dark magma, basalt." Sophia does the naming. "Let us talk to the members of the silicon, magnesium, aluminum, iron, and calcium atom families who produce this magma that has spread out from both sides of the crack in the ocean floor. This type of rock, once hardened, has paved the floors of the oceans, seas, and ponds." She greets some nearby atoms. "Greetings, my dear friends!"

One of the sulfur atoms immediately recognizes the Divine Beings. "Sophia, Elohim! So good to visit with you again. Where are Ruah and Dabar?"

"They are not present with us," Elohim answers. He begins some introductions. "Let me introduce you to some atom friends, Hydrojean, Nitro, and Phoson."

Once the mutual greetings are shared, Hydrojean inquires of this new friend, "Dear sulfur atom, how could you cocreate these new beings, the prokaryote bacteria, in such an intensely hot place, next to the molten basalt?"

"We had to cocreate differently than from other parts of our planet," the sulfur atom explains. "The red-hot magma pouring out of this ocean vent by the silicon, magnesium, aluminum, iron, and calcium atom families, along with the other five atom families, including your own atom families, were recycling all of the above particles to help us to cocreate." With pride, he goes on. "My sulfur atom family was crucial to our success. We were belched out of the underground magma with the other families. We had the ability to operate in very hot temperatures. We helped to supply the molecular compounds that could fuel and energize the new prokaryote bacteria beings that we cocreated with Ruah."

One of the phosphorus atoms spoke. "Yes, Ruah was crucial to the birth of this new being. She touched us deeply with her energy of love. The new being was vivified and came into life. Not only that, we have also been able to produce copies of this new being."

"I am sure then that you are familiar with RNA and DNA, those special molecules that have helped in the construction of these new prokaryote bacteria." Hydrojean wonders if they have been able to accomplish what she and her friends have done.

"Yes, we have," the phosphorus atom replies quickly. "Because of the hot environment here, we have had to cocreate these new beings to tolerate and thrive here."

"Your atom families have entered into the cocreative process beautifully." Sophia compliments them. She looks to the three atom visitors and directs her remarks quietly to them. "It is time to go forward in our journey." She turns back to their hosts. "Thank you for telling us about your cocreative work."

* * *

"You really transport us swiftly, dear Elohim." Phoson is dazzled by the rapidity of their journey. "I do notice that the atmosphere is becoming clearer. The continuous violent rain storms, the ponderous thunder, and the electrifying lightning have ended. The oceans, seas, and ponds are filled with waters. The waters on the crusts of the planet have continued to cut furrows into the surface and wash members of many atom families into the larger bodies of water to enrich them. The rains, lightning, and thunder are more periodic now than steady. For many days, the rains have been filling the oceans and seas and the many depressed crusted basins paved by the volcanic magma."

It is about October 12, 3.3 bya.

"Let us take a little detour." Elohim directs them down into the waters of the oceans. "Let's talk to some of the members of the atom families here. Maybe they can explain why the air is slowly clearing."

The travelers meet some calcium, carbon, and oxygen atoms. "Good to see you again, dear friends." The Divine Parent is joyous as he greets them. He introduces them to the three travelers accompanying Sophia and him.

"My dear Elohim, and you, Sophia! Good to see you again." A calcium atom is very overjoyed to see the Divine Beings.

"My atom friends would like to ask you a question," Elohim informs her.

"Yes, I have a question, dear friend." Phoson begins his inquiry. "We have noticed that the atmosphere is clearing. What is causing this?"

The calcium atom begins to give some explanation. "In recent days, I have noticed that many molecules made up of carbon and oxygen have been falling into the ocean. Our calcium family has communicated and mated with them to form a solid type of material here on the floor of the ocean. I am told by other family members that this is happening at the depths of all the oceans and seas. So possibly the atmosphere is clearing because the many, many gray gaseous molecules, made of carbon and oxygen, are being removed from the atmosphere. Thus, the removal of this molecular gas is clearing the air."

"Excellent reasoning, dear calcium atom." Sophia compliments her. "Let me give names to some of the elements. The solid material formed on the floor of the oceans and seas is named limestone.

A more generic name is rock. The latter is important for giving a solid floor upon which the waters of the planet rest."

"If you gaze off in the distance, you will see large, black formations of rock." In this underwater environment, the calcium atom points out its location. "At the base of these black rocks, there are cracks in the crust of this underwater world. The members of other atom families—silicon, iron, aluminum, and magnesium—which have been living in the hot magma under the crust of the planet, are being mated with other atoms, blown out through volcanic vents, and accreted to build these black rock formations on the floor of this watery ocean."

"I would like to give names to this black material and rock formation." Elohim names these two beings. "The former, you remember, is called basalt rock. The latter is named a mountain."

"Our gratitude, dear calcium atom, for your assistance to Phoson and his friends." Sophia delivers the note of gratitude. "We bid you farewell." Sophia turns to the atom trio. "Let's move from this underwater world to the shores of this ocean."

"What are those tall, circular, rocklike formations on the edges of the ocean?" Hydrojean looks intently at these strange objects. She addresses Sophia.

"Friend Hydrojean, let us again talk to some of the atom families that are a part of this object." Sophia wants Hydrojean to speak to these atoms.

"Greetings; my name is Hydrojean! I am a member of the hydrogen family." She initiates the conversation with an atom within this formation.

A phosphorus atom responds. "Always glad to meet new members of the atom family on this Earth."

"Friend, what is this strange home in which you live?" Hydrojean is delighted with the contact.

"You have not noticed, have you, that some of the members of the photon family reflect dangerous, deadly ultraviolet rays to our Earth?" The phosphorus atom questions her.

"No, I have not," Hydrojean says.

"My home is filled with various species of prokaryote bacteria. I am part of the exterior membrane of this being," he says. "Not too long after Ruah helped to energize our new being, some of our offspring decided to live on the surface waters of the seashore." He is sad and grief-stricken as he continues. "They were annihilated by the powerful photonic ultraviolet rays of our Sun."

"They could no longer live and grow?" Nitro was shocked by this revelation.

"That's right, friend." The phosphorus atom continues his tale. "This dreadful news came back to our community of bacteria. We still wished to live on the surface of the planet. So we devised a plan. We decided to build a home above the waters. We had to ask for volunteers from our prokaryote community to sacrifice their existence as members of a top layer. Exposed to deadly rays of our Sun, they were annihilated. But those of us who lived in the lower layers of this circular home were protected by those valiant bacteria on top. We built our layers from below. We have built a sizable multilayered home above the surface water. Of course, being near the shore, there are constant swells of the water that wash over us and bring us the nutrients we need to live, grow, and procreate new offspring."

"Can you give a name to this layered home?" Phoson asks Elohim.

"I name these bacterial homes, stromatolites." Elohim continues, "Notice that there are multitudinous numbers of stromatolites being built in various parts of the shallow ocean waters of planet Earth. They require salt water carrying nutrients for their bacterial community."

The phosphorus atom expresses a hope. "We are laboring vigorously to develop the ability to overcome our weakness in tolerating the Sun's deadly rays. We believe deeply that someday we will be able to adapt to these rays."

"May your community of atoms be successful, dear friend, in this cocreative mission." Sophia encourages them. "If you are successful, this may help some future beings who will come to live on this planet Earth."

"You may be prophetic, my dear Sophia," he says. After a short hesitation, the atom continues. "Thank you for visiting our stromatolite community."

"Good-bye, beloved prokaryote community." Sophia concludes the visit. "Let us proceed forward."

* * *

The two Divine Beings lead the atom trio forward. Again, they note the further clearing of the atmosphere. More radiation from the Sun now penetrates to the surface of the Earth. The radiation causes the water to evaporate and be stored as water vapor in the air. Storms, accompanied sometimes by thunder and lightning, later return some of this vapor in the form of rain to the Earth. Oceans and sea levels go up and down due to this process. Carbon dioxide continues to accompany the rain water as they return to the salty ocean waters. There, they mate with calcium to give birth to limestone rock on the ocean floors.

The waters near the shores of the planet are teeming with microscopic bacterial life. The prokaryote bacteria have been prolific in procreating offspring. The varieties of small microscopic bacteria increase greatly. The adaptations are written upon the DNA templates as the many changes and modifications occur.

The Divine Beings point out the progressive changes on the crust of the Earth. The volcanoes continue to produce more rock to expand the area of the crusts of the planet. When the rains come, they continue to wash atom material off the rocks. The waters cut deeper and deeper canyons in the rock formations. The rock sediments are made up of atoms that are carried by the ribbons of water and emptied into the ocean. Thus, the recycling process continues. Molecules of nutrients fill the oceans and seas to be used by the bacteria in cocreating newoffspring.

Finally, Sophia says, "Let's return to the present!" The time is October 22, 2.9 bya.

Chapter 40
Prokye 2 Develops an
Improved Replication System

"I still remember October 22, 2.9 bya, the day of our return from our journey back to the past!" Hydrojean recalls this epic adventure with delight.

"I was impressed by the cocreative work of all of the ninety-two atom families in the universe, the Milky Way galaxy, and our solar system." Phoson is enthralled by his vivid memories.

"To be a part of this stupendous creative journey into our universe is a deep privilege and a source of justifiable pride." Hydrojean adds to her joyous remembrance. There is a silence that intrudes their reminiscences. The memories seem to put them into a state of ecstasy. Finally, Hydrojean brings them out of their ecstatic trance. "Enough for the past. We need to concentrate on our present mission." She seems insistent.

* * *

The time presently is November 5, 2.5 bya.

"Right after we had our frightening adventure in invading that large prokaryote bacteria, I suggested that we try to develop an improved process for bacterial replication," Nitro recalls. "You all approved this suggestion."

Phoson quickly responds. "Yes, we did give our wholehearted approval." He pauses. "We undertook this mission immediately after our return from our astonishing journey into the past. We have experienced a frustrating twelve days (about 500 million years) in trying to improve our very primitive replication system."

"We have tried to improve the membrane openings." Hydrojean appears thwarted also. "We have assembled various molecules and amino acids into different structures, but none of them is an improvement over our present system."

"I believe we need a whole new concept." Nitro, the initiator of the present mission, says. "What can we develop that is revolutionary?"

"I have an intuition!" Hydrojean is excited. She continues with enthusiasm. "What if we developed a special structure within Prokye 2 where we can do replication?"

"What kind of a structure?" Nitro is inquisitive.

"A structure that can stretch. We can make it of long strands of proteins in an oblong, spherical shape." Hydrojean continues. "It will not have an encircling membrane. It will be composed of these fibrous protein strands with empty spaces in between. When the RNA messenger bases are created to transcribe the plan for the new replication, they will copy the construction plan from their DNA templates. The RNA messenger will then transfer this plan to a place inside the fibrous spindle. As the work of replication proceeds, the fibrous spindle will expand to accommodate the increasing size of the replication."

"Will there be any other parts of this structure?" Phoson questions her.

"We will need to construct a center point at two ends where the strands of fiber come together," Hydrojean responds. "They will be like small disks to which all of the fibers are attached."

"That reminds me of the imaginary poles that the Divine Beings described as being at each end of the solar planets we visited in our journey," Nitro recalls. He has a further question. "Will these be permanent structures, Hydrojean?"

"No. After our construction work is finished, we can dismantle them." Hydrojean is emphatic.

"That seems like a lot of extra work to do just to give us a structure within which to do our cocreative work," Phoson objects.

"My further intuition tells me that someday we may need to modify or change the operation of this structure, but for now I feel we need to create this new structure to help Prokye 2 in the replication process," Hydrojean prophesies.

"I have some of the same concerns as Phoson about your spherical structure." Nitro lets his opinion be known. He quickly adds, "Hydrojean, your creative intuition has led us to some wonderful developments." He then affirms her plan. "I am willing to follow your intuition and work at developing this structure."

"I concur." Phoson adds his affirmation. "Let's commence our work."

This trio, with the help of the other five atom family members, begins to construct the fibrous strands from various molecules, proteins, and other raw materials available there. Nitro and Phoson lead the effort at the opposite poles of this spherical structure to fabricate the disk-shaped center points that that anchor the fiber lines.

After they have completed this oblong, spherical structure, the laborers, led by the atom trio, stand aside to admire their creation.

Ruah arrives on the scene. "A magnificent accomplishment, my dear friends. You have outdone all of your previous accomplishments." His gaze is on Hydrojean. "Your feminine intuition is uncanny!"

"How true, Ruah," Phoson says. "Nitro and I had some reservations because it is not to be a permanent structure. It seems like extra work to build and then dismantle it."

"Remember that new creations can be changed and mutated as your cocreative work continues." Ruah adds some insight. "You may discover some added functions of this structure as you modify it. This is a good start."

"Ruah, can you give names to the spindle fibers and to the disklike center points at the poles of this sphere?" Hydrojean requests.

"We will call the long, expandable, hollow protein fibrous strands, microtubles." Ruah begins her naming. "The porous spherical structure made up of these microtubles will be called the spindle apparatus. Finally, the disklike protein center points at the opposite poles of the spindle apparatus that bind these strands will be called the kinetochore."

Nitro remembers. "We still need to take our construction plans and record them on our DNA template so that all further reproductions of Prokye 2 will have this new structure as a part of itself."

They construct the combinations of RNA nucleotide bases utilizing some free-floating RNA nucleotide bases to create an RNA messenger with the encoded design plan for the new spindle apparatus. Then it travels to the site of Prokye 2's DNA home and records these innovations in the appropriate place on the DNA template.

Having finished this mission, they discuss the future. Nitro has a plan of action. "I believe that our development of a rotating tail and our spindle apparatus can help some of the members of the prokaryote family and us to move forward in our joint co-creative endeavor." He looks at Hydrojean and Phoson intently and continues. "I believe we need to invade another larger prokaryote bacterium and offer our developments to them."

"We tried that before and were ordered out of their domain. We seemed to pose a threat to them." Hydrojean is very concerned about such a venture. "I am fearful of trying this strategy again."

Phoson is insistent. "I am fearful also, Hydrojean. For the same reason. I feel we should follow Nitro's plan. We have these new developments. They could be helpful in pushing forward the universe's mission to continue its evolving creativity."

"I guess you are right, Phoson." Hydrojean seem reflective. "When I think of how the whole universe is cocreating, we need to put our fears aside and move boldly into the future."

"That's following the creative spirit." Ruah commends her.

"Prokye 2, find another larger prokaryote for us to enter and tend an offer." Nitro's voice is commanding.

Chapter 41
Prokye 2's Surprise Reunion
with Prokye 1

"Prokye 2, find the largest prokaryote bacterium that you can!" Nitro recalls their previous visit to a large bacterium. "Remember that our last incursion into a prokaryote bacterium was disastrous since there was not sufficient nutrition for it and us. We had to evacuate it promptly."

The time is November 6, 2.4 bya.

Prokye 2, propelled by its rotating tail, is able to be choosey in its selection. After quite a journey, Prokye 2 spots a possible target for entry.

The atom trio watches Prokye 2's maneuvering into position to make the entry. "A good choice, Prokye 2!" Hydrojean shouts. "I think we will be able to establish a home there."

"We have to be cautious in our approach." Phoson also remembers. "They may be very frightened by our visit and request." He looks to Nitro and says, "Nitro, you be our spokesperson, okay?"

"I will do the best I can, Phoson." Nitro is agreeable.

"I affirm you as our leader, Nitro. I trust you deeply to help us find a new home where we will be able to share our creativeness with our host." Hydrojean is supportive. "I believe we have a lot to offer them with our rotating tail assembly and our spindle apparatus."

Assisted greatly by its undulating tail, Prokye 2 maneuvers through an opening in this very large prokaryote bacterium and guides itself to the middle of the latter's interior.

Some of the members of the atom families are busy working on a new creation. They look up, surprised and bewildered by this bold incursion into their space. Fear begins to overtake them. One of the carbon atoms assumes the responsibility to question these intruders. "Who are you? Friend or foe?" There is a sense of command, yet fear, in his voice.

"We are friends!" Calmly, Nitro answers. "We come to offer and share with you some of the new creations we have developed since our host, Prokye 2, came into existence on October 9, 3.4 bya."

Quarko has been listening carefully to the initial words of the visitor. He is searching his memory to identify the latter's voice. Suddenly, his face lights up. He blurts out, "You old fool!" He pauses. "If it is not notorious Nitro! From where did you come, old friend?"

"It is Nitro!" Quarkie shouts. "I also see Hydrojean. I don't recognize their friend."

"What a wonderful surprise!" Hydrojean is overjoyed. "Dear Quarkie, I despaired of ever seeing you again. Oh happy day!" She turns toward her friend, Phoson. Then she says, "Dear quark trio, this is Phoson. We met him shortly after Prokye 2 launched us as a new prokaryote bacterium. We have worked with the other five atom families to carry out the cocreative work requested by Ruah."

"I am filled with joy at our reunion." Quarkoff extends a hearty greeting. "I am happy to meet any friends of Nitro and Hydrojean."

Nitro is bewildered by this happy turn of events. After their initial experience, he had expected difficulty in Prokye 2 gaining admission to any larger prokaryote bacterium. Here they are welcomed with open arms, and by friends who have known each other before. He says, "I guess I should not be surprised, since I have previously met former atom friends during my journeys. It was so important to be accepted and hosted by a larger bacterium so we can share our creative innovations. We can never have received a better welcome! Thank you!"

"You have spoken of new creations and creative innovations." Quarkoff begins to quiz his long-lost friends and new friend. "What are they?"

"We have developed a tail assembly with a rotating, flagellating tail," Nitro begins. "We have continued to perfect its operation. It gives Prokye 2 a big advantage. The tail is connected to its exterior membrane wall. It uses power from Prokye 2's energy system to rotate. Hydrojean got the idea of the tail from observing the tail of the icy comets we had observed crashing into planet Earth. This innovation helps Prokye move about its environment, especially in search of nutrition."

Quarko challenges them. "Knowing you, Hydrojean and Nitro, and your sloppy habits, I would bet that your innovation will be of no use to us because you failed to encode the design plans on your DNA template. You probably would have a hard time remembering how to reconstruct this tail assembly."

"Still the provocateur, nasty Quarko." Hydrojean is quick to respond with sarcasm. "Let me assure you, quirky Quarko, that we had an RNA messenger carefully record our innovation upon our DNA template."

"Despite your nastiness, we are still willing to share our creative tail assembly with you." Nitro is caustic to Quarko.

"Phoson, you need to be warned about our nasty friend's high jinks." Hydrojean gives him a warning. "At times, he is a questionable friend. At other times, he can be a quintessential quark to all of us." She looks at him severely and continues. "I thought you were going to reform your ways, nasty Quarko."

"You have always taken me in the wrong way, darling Hydrojean." Quarko's remark is dripping with sarcasm. "The Divine Beings allow me to be my humorous, quirky self."

"They just get frustrated by your spirit of divisiveness," Hydrojean concludes.

Phoson takes up the conversation. "Dear Quarkie, Quarko, and Quarkoff, we have one other innovation, a spindle apparatus. This is a building made up of expandable hollow strands of protein, called microtubules, that are bent in a spherical shape. It is held in place by two disklike center points, called kinetochore, at opposite poles of the structure. Within this structure, we do our replications of new prokaryote bacteria. When finished, we disassemble it to wait for the next time of reproduction."

"What kind of an improvement is that?" Quarko continues his criticism and badgering. "You could do your replications just as well without such a monstrosity."

"Ruah called it a 'good start.'" Phoson stands up to Quarko's put-down. "She felt that as time goes forward, this structure could mutate and fulfill future functions of the evolving prokaryote bacteria community. We are satisfied that this is a worthwhile structure." He adds, "Thank Hydrojean and her intuition for dreaming up this spherical framework."

"Idiotic intuition, I would call her creativity," Quarko calmly retorts.

"I note a bit of jealousy, queer Quarko," Hydrojean lashes back at him. Then, after a short lull, she changes the subject. "Quarkie, what new creations have you developed in Prokye?"

"I intuited the need for Prokye 1 to grow much larger." Quarkie recounts her intuition.

"Feminine intuition is uncanny, is it not, Quarkie?" Hydrojean applauds their feminine gift.

"I could be nasty." It's quirky Quarko spouting off again. "I will modify my behavior to keep peace."

"Thank you for your graciousness." Hydrojean is surprised by her quark mate's control. She returns to ask Quarkie, "Why did you decide on a larger size for Prokye 1?"

Quarkoff takes up the conversation instead of Quarkie. "I had suggested that since we were much larger, we would be able to host smaller bacteria. I had no idea what the latter could contribute to us. I thought we could offer them hospitality."

"It is strange," Quarkie intrudes. "When Quarkoff made this suggestion, I was fearful of what the presence of a guest bacterium would do to us. In fact, Quarko was fearful also."

"I was not fearful, dear Quarkie." Quarko does not like this insinuation of fear on his part. "I just thought it was a stupid idea. Period."

"You, Hydrojean, Nitro, and Phoson show up as guests and offer us some new creations that we can integrate into Prokye 1." Quarkie is amazed at the turn of events. The arrival of guests who are friends has dispelled all fear of a possible catastrophe.

"Do not forget my innovation, dear friends." Quarko puffs up with pride. He continues. "I developed the idea of enclosing the DNA template with a membrane wall since the latter is so important to our creative activity." He puffs up more. "Let me show you this beautiful membrane that surrounds our DNA template. Ruah gave a name to the wall and its contents. She called it a nucleus since it is similar in the home where we quarks, who make up the protons and neutrons in an atom, reside."

"Congratulations, quintessential Quarko." Nitro is complimentary of his sometime adversary. "The nucleus is a beautiful name and a marvelous creation."

"I might add that I do not want to appear to be too proud." Quarkoff speaks of an additional innovation. "Once this membrane enclosed the DNA template, I suggested that we reorganize the combinations of DNA nucleotide bases, which have the complete plan for Prokye 1 encoded there. My suggestion was that we organize the DNA bases in the form of a spiral helix. I thought this would more efficiently utilize the space and facilitate the work of the RNA messengers to copy parts or the whole plan of Prokye 1."

"Nifty Quarkoff demonstrated great creative skills, I must admit." Unbelievably, Quarko compliments his quark mate. There is a pause in their joyous conversation.

Phoson has been observing the reunion of these atom friends. He has a question to pose. "So we have become your guest. We need to attempt to carry out the mission that we came to accomplish. Namely, we would like to share our creations with you. By living inside of you in this semi-aqueous environment, we are sharing in your innovations. We need to transfer our innovations to Prokye 1."

"I never thought of this," Hydrojean muses. "When we become your guest and you the host, we lose our identity as a prokaryote bacterium and become just a part of you." She mulls over her last comment. "I do not feel bad about this loss. Our loss is a big gain for all of us and the creative process of the prokaryote bacteria family."

"A wonderful insight, lovely Hydrojean." Quarkoff accepts the new living situation. "Our reunion together again in Prokye 1 is a true gift to ourselves and the evolving process of our planet Earth and the universe."

"There is still an important work to accomplish," Phoson suggests. "If you are to share in our creative developments, we must build these two innovations into Prokye 1 and then transfer our construction plans for the tail assembly and the spindle apparatus, microtubules, and kinetochores to the DNA template and helix within the nucleus of Prokye 1."

"Let us do that right now." Quarkoff assumes leadership for this work. "Let us jointly construct an RNA messenger to do the initial transcription."

Quickly, the six shipmates begin to construct the combinations of RNA nucleotide bases to create this RNA messenger. It enters Prokye 2's DNA template home, plugs into and copies its design plans for these two new creations. The sextet works from the plans of the RNA messenger. They assemble the combinations of amino acids to build these proteins into the flexible, tapered tail. They attach it to Prokye 1's outer wall. Then they set out to build the spindle apparatus, including the microtubules and kinetochores, again using the combination of amino acids and other raw materials available in Prokye 1's semi-fluid interior. They connect these new functions to Prokye 1's power plant.

The quark and atom builders then direct the RNA messenger to travel through an opening in the membrane of Prokye 1's nucleus. It enters the DNA template helix and encodes the plan for these new creations. Mission accomplished.

Quarkoff asks Prokye 1 to order its power plant to send some energy to allow the tail assembly to rotate. I notice that the tail is enlivened. It begins slowly to rotate. They can feel Prokye 1 moving in its exterior watery environment, powered by its rotating tail.

"I sense we are moving impelled by your creation, the tail assembly," Quarkie shouts to their friends. "Congratulations!"

"You have been successful in your mission of sharing your new creations with us." Quarkoff is complimentary of his friends from Prokye 2. "Our fears of hosting a guest within our larger bacteria were groundless. You have outfitted Prokye 1 with great gifts."

"We have completed our assembly in record time," Quarko boasts. "I still seriously question whether your spindle apparatus is a development or just a meaningless relic."

"We will find out sooner or later," Quarkoff counters Quarko. "You thought that any attempt to have a guest bacteria occupy our spacious quarters would end up in disaster." He pauses and smirks. "How wrong you were, queasy Quarko!"

"Let us let time decide who the genius is and who the fool is," Quarko calmly responds.

"Why do we not work together now within the spindle apparatus to construct a mirror image of Prokye 1, but with these new innovations built into its structure," Quarkoff suggests.

"Talk of a replication reminds us that we are losing a dear friend." Hydrojean is concerned.

"Whose friendship are we losing?" Phoson asks.

"Now that we have transferred our DNA covering the two creations to Prokye 1's nucleus, Prokye 2 loses its existence," Hydrojean responds. "Yes, our bacterial home now is but a part of our host."

Hydrojean, Nitro, and Phoson gather with the other five atom families inside of Prokye 2 to bid a fond farewell to their friend. "Thank you, Prokye 2. You have been a marvelous host to us. Good-bye."

Prokye 2 tells them it is resigned to its fate. It has carried out its mission! It has accomplished its purpose in creation history. The atom friends admire its acceptance and courage.

"Prokye 2, Hydrojean, Phoson, and I have enjoyed our close friendship with you. We are truly grateful to you!" Nitro comments on their relationship. "Together, we have experienced heights of great joy, times of great danger, and deep disappointment. You have always worked with us diligently and faithfully." He pauses, then continues. "Just as Hydrojean prophesied that our new spindle apparatus may have a much greater future than we envision presently, I believe that bacteria of the future may someday be able to establish more permanent friendships. I foresee a time when the interpenetration of a smaller bacterium into a larger one will lead to a special, lasting friendship, each retaining its own individuality. They will work together for a higher purpose, I prophesy." He ends. "We will have to await future creative developments."

The other five members add their sincere gratitude to Prokye 2.

They begin to dismantle Prokye 2 in order to use its molecules, compounds, atoms, and other particles in the construction of new prokaryote bacteria.

Then the six friends, along with the five atom families inside of Prokye 1, cooperate to begin the construction of a replication with the two improvements contributed by their guest. All of this is taking place within the spindle apparatus.

Chapter 42
Prokye 3 Comes Home

"We have continued to improve our power system since October 24, 2.8 bya," Oxydon recalls. "What is our next mission?"

Nitrojoan mulls over Oxydon's question. Temporarily, there is silence among the atom crew members of Prokye 3. Then she comments, "Ruah has told us that her energy of love is always present, at a profound level, to enliven our cocreativity. I experience her energy now." She pauses. "I believe this is the momentous time to make a decision about our future course of action."

The time is November 12, 2.0 bya.

"What can we do? What new work of cocreation can we undertake?" Oxydon is stymied.

"Prokye 3, our personal host, does have an improved power system. We have developed ATP as a carrier of this energy to all its functions," Nitrojoan recalls.

"How do we share our creative developments with other prokaryote bacteria?" Oxydon is still questioning. "We do produce bacterial replicas of Prokye 3 periodically."

Photie has been present to this conversation, but silent up to this time. He says very calmly, "Dame Fate advises us to devise a perilous plan."

"Perilous plan? What do you mean, Photie?" Nitrojoan is shocked.

"I do not like the word *perilous*, Photie." Oxydon seems fearful. "I do not want to go through a situation similar to the terror and threat of annihilation we encountered when that enormous prokaryote bacterium swallowed us on October 30, 2.5 bya."

"*Perilous plan* means that we need to take a rambunctious risk." Photie seems charismatically confident.

"What kind of a 'rambunctious risk?'" Nitrojoan is still anxious about Photie's "perilous plan."

"Our profound power can only be shared with a plump, portly prokaryote bacterium." Photie displays unusual firmness.

"After our experience with the demonic leader in our last excursion inside of a large prokaryote, that is a 'perilous plan.'" Nitrojoan expresses her deep concerns. "Your plan scares me greatly, dear Photie."

"I have the same reservations, Photie." Oxydon affirms Nitrojoan's fears. He pauses to reflect on the whole conversation. Then he continues in a surprising manner. "Since you delivered Prokye 3 and us out of that dangerous event with that cruel leader, I have great respect for your ideas, and especially, your captivating courage." He again pauses. "So ... so I am ready to follow your leadership. Let us take that 'rambunctious risk' with you leading us."

Nitrojoan is very shocked by this turn of events. From supporting her fears, Oxydon has moved to supporting Photie's dangerous plan. "I am full of anxiety and conflict." She expresses her concerns.

"Nitrojoan, I can feel for you." Oxydon empathizes with her. "Yet, we would not be here if it was not for Photie's courageous leadership. I am ready to put aside my own fears to follow his risky plan." Moments go by in silence. Then he says, "I hope you can join us in carrying out this 'perilous plan.'"

Nitrojoan is in deep thought for several moments as Photie and Oxydon await her response. Finally, she utters slowly, "It is because of my deep confidence in you, Photie, that I join you in this 'perilous plan.'"

"Thank you, thank you, noble Nitrojoan and observant Oxydon, for your succulent support." Photie is truly grateful to his friends. "Precocious Prokye 3 needs to guide us into our new prodigious prokaryote port of call."

Oxydon gives Photie's order to their host. Prokye 3 moves forward to the nearest large prokaryote bacterium. Carefully, it steers itself through one of the large openings on the membrane wall of this big being. They come to rest in the middle of its aqueous environment. Photie appears resolute and confident. His two friends have residual fears and deep concerns as to how they will be greeted.

"Fabulous friends!" Photie peacefully shouts out his greetings. "We come with an inventive innovation of profound power. Sumptuous sharing of our cunning creativity is our glorious goal."

There is a momentary pause. Suddenly, the quark trio, as one, shouts in joy and exultation. "It's fabulous Photie!"

Hydrojean and Phosie excitedly move toward their long-lost friend. "Photie! Photie! How good to see you again."

Oxydon and Nitrojoan are bewildered, yet brightened by the overwhelming greeting by the atom members of their new bacterial host. In the meantime, Phoson, who is a more recent member of

Prokye 1's crew, is amazed that another reunion of friends is taking place.

"Dame Fate has arranged this raucous reunion, dear fancy friends." Photie is overwhelmed by the greeting. The tumultuous, joy-filled scene calms. Photie turns to his shipmates on Prokye 3 and asks them to come forward. "Fabulous friends, let me introduce two sweet, supportive shipmates of mine, Nitrojoan, a merry member of the nifty nitrogen family, and Oxydon, a magnificent member of the outstanding oxygen family." They exchange greetings with all of Photie's friends.

Hydrojean then introduces one of her former crew mates on the former Prokye 2. "Photie and friends, this is Phoson, a member of the phosphorus atom family." There is a sincere welcome given to Phoson by all present.

Quarkie asks their new guests, "Photie, you mentioned that you arrived here with an 'inventive innovation,' a plump, portly power system. Could you tell us more?"

"Omniscient Oxydon, please tell quiet Quarkie and her gentle friends about our cunning creativity." Photie turns the group's attention to his crew mate.

"Dear friends of Photie and new friends, we decided on a mission after consulting with Ruah." Oxydon begins his story. "Ruah did not decide on our specific mission, but she encouraged us. We were and are always aware of her love energy impelling us forward. As a group, we chose to create a more powerful and efficient energy system. We developed a process to develop a new power source. About October 24, 2.8 bya, working with the other atom families, we constructed a compound made up of a phosphorus group, a sugar group, and a special group, adenine, a nucleotide. Ruah named this compound adenosine triphosphate and shortened it to ATP. We tried it out and it worked extremely well. We have come to visit you to share our cocreative development with you."

"Congratulations to the three of you and all of your atom family friends in Prokye 3." Quarkie is quick to applaud their efforts. Addressing Photie, she continues. "I remember when you volunteered to be part of the crew of Prokye 3, Photie. You addressed Ruah: 'I volunteer to be an ambitious apostle of courageous cocreation.' You prophesied that 'dazzling destiny' had summoned you. You have shown ingenuity, courage, and responsibility in accomplishing your 'dazzling destiny.'"

"Lovely Quarkie, I thank you, thank you for your wondrous words of passionate praise." Photie displays sincere shyness.

Nitrojoan adds to his adulation. "Photie is truly courageous. He delivered us from the clutches of a demonic leader of a large prokaryote bacterium, which swallowed up Prokye 3. He threatened us with annihilation. Photie stood up to him by lively, wise alliteration and delivered us from destruction."

Oxydon then took up the tale. "Photie showed his remarkable bravery in urging us to enter into your large home. Remembering the previous experience, we were reluctant to take a chance and enter here. Trusting Photie in our first harrowing experience with a large prokaryote bacterium, we decided to trust him in leading us here." He turns to Photie and says, "Fantastic Photie, Nitrojoan, our whole crew of atom families and I are deeply grateful to you."

Photie is speechless. All of his friends applaud him. He bows humbly in gratitude. Then he addresses all present. "Quarko, my funny friend, what dazzling destiny have you attained?"

Uncharacteristically, Quarko decides not to respond to his "funny friend" title. He addresses his question. "Quarkie, our resident feminist, intuited that we work at vastly increasing the size of Prokye 1. In addition, at my ingenious suggestion, we have placed a double membrane wall around our DNA template." He is hesitant, deliberating whether to recount the final part of their creative mission. He decides to continue. "Quizzical Quarkoff brilliantly proposed that we reorganize the DNA nucleotide bases into the form of a spiral helix in order to allow for more efficient storage of the bases and for RNA messengers more easily to copy all or part of the DNA template. So all three of us contributed to these creative innovations to Prokye 1."

Nitrojoan asks the members of the former Prokye 2 what mission they had accomplished. Nitro takes the lead and says, "I am so happy to become a friend with another member of our nitrogen atom family, Nitrojoan." After his personal note, he continues. "We worked on developing a rotating tail assembly, which we attached on the exterior of Prokye 2. This innovation allowed us to be able to move around our watery environment much more easily. Then we worked to develop the expandable spindle apparatus within which we carry out the construction of the replicas of Prokye 2."

Right at this moment, Dabar and Ruah arrive on the scene. Dabar greets them. "Ruah and I have been observing the reunions of Prokye 2 and Prokye 3. We must congratulate all of you for missions accomplished and, as Photie so well put it, 'dazzling destiny' vigorously pursued."

"We interrupted your conversation. You have some tasks to perform." Ruah is sensitive to their needs.

Nitrojoan makes a suggestion. "Since we come to share our innovation with you, where do you suggest, my dear quark friends, that we place our new power system?"

"I think it should be located close to our own power system." Quarkoff is quick to respond.

Hydrojean brings a somber note into the discussion. "When you replace our new power system from Prokye 3, the latter ceases to exist."

"Prokye 3 no longer lives and exists? That is not fair!" Oxydon is upset. He turns to Dabar in his surprise and anger. "Dabar, why does Prokye 3 lose its existence?"

Phoson is touched by Oxydon's expression of impending grief and loss. "I felt that same way as Oxydon when we realized that, after we transferred our DNA bases containing our new creations to Prokye 1's DNA helix and we constructed its innovations into Prokye 1, Prokye 2 would lose its existence." He looks at Dabar too and says, "Why?"

Dabar gathers the nine atom characters. Ruah's love energy is palpable. "You who are members of atoms families are indeed fortunate. Once created in this universe, you continue to live on. You, the quark trio, have lived in many forms. First as individual quarks, then in the proton of the first hydrogen atom. Next, you were part of a helium 4 nuclei. Later, in the first supernova explosion, you became a carbon atom. Finally, you became a part of a hydrogen cyanide molecule. Think of your additional roles in the formation and growth of living beings like Prokye 1. You will continue to exist as long as this universe." Dabar pauses and his gaze is more intent. "There will be created beings in the universe that will die. You have seen suns die. Further, possibly you can develop a branch of bacteria that will be able to survive and relate to other larger prokaryote bacteria. This will require some creative effort on your parts."

"What does it mean to die?" Nitro asks.

"To die is to stop living as a being." Dabar goes on. "So Prokye 3 will die once it loses a vital part of its internal functions and transfers its DNA bases containing the design for its power plant to Prokye 1's DNA helix. Once Prokye 3 gives up its DNA, it dies." Dabar reminds them, "All of the atom members, including some of you, amino acids, RNA and DNA nucleotide bases, proteins, and other substances will continue to live. They and you are freed to continue further creative work within and outside of Prokye 1."

"So we are very fortunate to continue to live in the universe and continue to carry out our creative mission." Hydrojean is expressing her gratitude.

Dabar asks. "What was Prokye 2's attitude toward its death?"

"It was resigned to its fate." Nitro continues. "It was happy that it had contributed to the creative process of the universe."

"I believe you will discover that Prokye 3 will be resigned to its fate also," Dabar comments. "When death once occurs, it becomes the law of the universe for some living beings. You will have to get used to the event of death, difficult as it may be to lose a dear friend."

"Let us get on with the transfer of Prokye 3's DNA for the construction of this new power system." Oxydon is anxious to begin.

Photie politely addresses him. "Outstanding Oxydon, ingenious intuition tells me that we should not dismantle fantastic Photie 3's puissant power system after we transfer its DNA template to our hospitable host." The rest of the atom members focus intently on Photie. "Once transferred, I believe that we should build a memorable, majestic membrane wall around this important, potent power source and preserve its dynamic DNA within this wealthy wall."

Nitrojoan has absorbed Photie's plan. "I would advise all of you, my friends, not to discard Photie's 'innovative intuition.'" She directs her next comment to Ruah. "I believe deeply that your

289

energy of love works exceedingly well within Photie. I have been gifted by his wisdom delivered so simply and timely during our journeys together in Prokye 3."

Quarkie gives her assent to his plan. "Yes, I believe Photie's 'innovative intuition' should be followed. The nine members come to a quick consensus on the plan.

Sadly, they bid their farewells to Prokye 3. They each express their gratitude to it for its creative work.

Their grief expressed, the nine friends initiate the building of a double wall membrane around Prokye 3's power plant. They make a special place within this wall for the DNA template for this system. Then they cooperate and construct the RNA nucleotide bases from the raw materials present in Prokye 3's semi-aqueous interior into an RNA messenger. The latter enters Prokye 3's DNA template to begin the transcription of its power-system plan. The RNA messenger first builds a special DNA template for this plan within its membrane-walled power plant. (This completes the fulfillment of Photie's "ingenious intuition.") Only then does the RNA messenger transport itself through Prokye 1's nucleus's membrane wall guarding its DNA helix. It opens up its DNA template and at a special point on the helix encodes the exact design plan of Prokye 3's power plant.

In the meantime, Nitrojoan, Oxydon, and Photie carefully dismantle the old membrane wall of the former Prokye 3. This leads to its demise, but its great contribution pushes forward the creative process of the Earth and the universe.

Oxydon asks Ruah, "What should we call this new power source bounded by a strong membrane wall, still possessing its own DNA helix?"

"I name this new power system, where the energy ATP is generated, the mitochondria," Ruah proclaims.

"Mighty mitochondria!" Photie repeats the name softly. "A potent powerhouse!"

"Let us disconnect the old power system and put the new mitochondria system into operation," Quarko orders. "Let us find out if it really works. I have my doubts that fatuous Photie's intuition will work."

"You are the stupid one, Quarko." Quarkie is incensed. "You are trying to pay him back because he called you 'funny Quarko.'"

Photie is unfazed by the comment. "Thank you, thank you, for your caring compliment."

Dabar intrudes himself into the dispute. "Quarkie, ask Prokye 1 to turn on the new power system that replaces the old one, now that it is in place."

Quarkie makes the request. Prokye 1 powers it up. The ATPs flow to all parts of its being.

"What an improvement!" Quarkoff shouts. "It is much more powerful than the old system."

"Your 'innovative intuition' is a great addition to Prokye 1. Thank you, dear friends from the former Prokye 3," Quarkie exults.

Uncharacteristically, Quarko seems embarrassed after his challenge was taken up successfully. "I just wanted to find out if it worked."

"What is our next mission, Dabar and Ruah?" Quarkie inquires.

"That is for you to decide, my dearest atom friends," Dabar advises them.

"Remember that I am with you always as you decide and carry out your cocreative work to further the evolving of our Earth and our universe," Ruah adds.

"Why do you not just rest for a time?" Dabar says. "Enjoy your renovated home before you begin the next phase of your journey."

They share the stories of their adventures in the universe and on the Earth. They rest.

Chapter 43
Homecoming for the Crew of Prokye 4

"We need to find a new host in order to share our creative development with them." Hydrojohn urges his shipmates on Prokye 4. "We have been traveling for almost sixteen days, October 30 to November 15."

The time is November 15, 1.9 bya.

"We have been successful in making reproductions of Prokye 4, but we need to share our innovations with a bigger prokaryote." Phosie concurs with him.

"Let us request Prokye 4 to find a very large prokaryote bacterium," Oxyjoy says. "Maybe we can be as successful as we were on our first visit to a large host."

Phosie takes the lead. "Prokye 4, please find a much larger prokaryote bacterium and make an entry. I will handle the introductions."

I watch Prokye 4 find a suitable host. It makes an easy entry through one of its membrane holes. It comes to a halt in the middle of this semi-aqueous environment.

"My friends, we come to help you. We pose no threat to your existence." Phosie is forceful in her greeting but respectful. "We have come to share our new creation."

Quarkoff is the first to respond. He is wary but senses that he knows the speaker. There is a pause. Then his memory tells him. "If it is not our long-lost friend Phosie."

"It is Phosie!" Quarkie recognizes her. "How wonderful!"

Oxyjoy and Hydrojohn had not expected such a surprise welcome. Phosie turns to them and says, "This is a homecoming for me. I was launched from Prokye 1 on October 16, 3.2 bya. We are reunited on November 15, 1.9 bya, about 1 billion years later. What a long, long time!" Then she moves forward to greet her long-lost atom friends.

They begin their reunion by sharing stories of their adventures of the past month. Having completed this time of storytelling, Hydrojean asks the three visitors, "What innovations do you have to share with us? We have already learned of the contributions of the tail assembly and spindle apparatus from the crew of Prokye 2 and the mitochondria from the shipmates on Prokye 3. We await eagerly your new creativity."

"Consequential creativity has come forth from you, my fanciful friends, I am positive!" Photie is affirming the newly arrived guests.

Oxyjoy begins his story. "Phosie, Hydrojohn, and I observed that prokaryote bacteria could not live on the surface of the seas and oceans. The strong ultraviolet rays would snuff out their existence. We decided to devise some way to harness these rays from our Sun and turn them into energy. We developed an aldehyde molecule platform, which captures the violet-blue rays from our Sun, and a methyl molecule platform, which captures the red rays. When these rays strike these receptors, an electron is released that carries the Sun's energy to Prokye 4. This energy can also be used to provide nutrition for us. The energy carried by the electrons synthesizes the carbon dioxide molecules and water to produce sugar. The byproducts are two atoms of oxygen. We release my family members as waste. At first, this was repugnant to me. Later, I intuited that someday these waste atoms will fulfill a great service to the cocreative process. The future will tell us."

"We would like to transfer our DNA template for this innovation from Prokye 4 and add it to Prokye 1's newly developed DNA template and helix," Hydrojohn proposes.

"Do you realize that Prokye 4 will die due to this transfer?" Hydrojean is blunt.

"What does it mean to die?" Hydrojohn asks.

"Dabar said that to die is to give up living as a prokaryote being." Hydrojean utters this with gentleness. Then she adds, "You, atom family members living within Prokye 4, will continue to live."

"Oh, I understand what you mean." Hydrojohn recalls some memories. "Oxyjoy and I have observed death already. We have seen prokaryote bacteria die when exposed to the rays of our Sun. Some members of the prokaryote family live in large mats above the waters near the shores of the ocean. Some of their members living on the top layer die to shield others underneath from our Sun's deadly rays. So this is what it means to die."

These bacterial mats jog Phoson's memory. "When Elohim and Sophia took us back to the past, we saw these mats. Elohim gave them the name stromatolites."

"What does back to the past mean, my friend?" Oxyjoy questions Phoson.

"We will discuss that later." Phoson avoids the subject.

Phosie changes the subject. "We have a more immediate task at hand. We need to transfer our DNA template encoded with our new creations by way of an RNA messenger to Prokye 1's DNA template and helix. First, let us build an RNA messenger made of nucleotide bases from our free-floating raw materials and RNA bases." She initiates the process of transfer by joining Oxyjoy and Hydrojohn in the construction process.

I watch them construct the RNA messenger. Then the latter enters Prokye 4's DNA template home and copies its design plan for the sun ray receptors. Then it proceeds to invade Prokye 1's nucleus membrane and search out the latter's DNA template and helix. It encodes this design plan there. Now Prokye 1 has the ability through its DNA template to pass on this innovation to its entire future offspring.

"Mission accomplished!" Phosie exalts. Her two shipmates share her enthusiasm. They have one more poignant responsibility to fulfill. Phosie initiates this important duty. "Thank you very, very much, Prokye 4, for being such a wonderful host to the three of us. You have contributed mightily to our creative mission."

"It has been a privilege to serve in your creative mission and to share a friendship with the three of you and all of the atom members that created me and sustained me all these many, many years," Prokye 4 responds. "I feel a sense of deep fulfillment. So I am ready to cease my existence among you. I too am very grateful to all of you."

With heavy hearts, Phosie, Hydrojohn, and Oxyjoy utter with one voice, "Farewell, beloved Prokye 4!"

This responsibility discharged, they dismantle its membrane wall. This frees its entire atom family members making up the wall and all atoms living inside to find other creative work either within Prokye 1 or in the watery world outside of this prokaryote.

The Divine Beings arrive on the scene. Sophia addresses the twelve atom characters. "Another reunion has taken place. I observe that Phosie, Oxyjoy, and Hydrojohn have also found a home in Prokye 1. So your reunion is complete."

"When some of you left Prokye 1 back in October, I mentioned that only the future would answer your question as to whether or not you would meet your crew mates again." Ruah recalls her past statement. "Well, the future has arrived and you are all reunited, along with some new friends you have met on your cocreative journeying."

"We were the last group to come back to Prokye 1, our original home," Phosie says. "We contributed a new process for utilizing the power of our Sun's ultraviolet rays. We need you, the Divine Beings, to give a name to our process."

Sophia immediately responds. "Since your new innovation captures light from the Sun, let us call this process photosynthesis. The place where these molecular receptors reside will be called the chloroplast."

"I like this name." Oxyjoy is happy. "I prophesy that photosynthesis will have a very important place in furthering the cocreation of the future Earth. I cannot say exactly how or when. It is destined to contribute mightily."

"Let us allow the future to determine if your prophecy will come true," Elohim responds. "We will just have to wait and see."

"What is our next mission, Elohim?" Quarkoff adds.

"You must determine that." Elohim turns the question back to them.

"I suggest that we just work at perfecting and coordinating all of the new functions we have established within Prokye 1," Quarko opines. "This is not an intuition but merely a suggestion."

"Suggestion or intuition, I agree with Quarko's proposal." Phoson supports him.

As they discuss this suggestion, a consensus arises about their next mission.

"Some of you have not experienced journeying back to the past," Ruah comments. "I took the quark trio on the first journey. Elohim and Sophia took Hydrojean, Nitro, and Phoson on the second journey. I propose that Dabar lead a back-to-the-past journey for Photie, Oxydon, Nitrojoan, Phosie, Oxyjoy, and Hydrojohn."

"What does that mean?" Oxyjoy questions her.

Quarkoff begins to describe such a journey. "My friends, we have been living in the microcosm of the Earth under the waters of the seashore. We have been out of touch with what the other ninety-two atom families are cocreating in this vast universe, above, on, and below the surface of the Earth. I have been amazed at the co-creative work being accomplished by all of the atom families in our magnificent universe and planet Earth. Of course, Ruah energizes and encourages them, as she has done with all of you."

"So you, Ruah, propose that the six of us accompany Dabar on such a journey into past time." Oxyjoy understands the nature of the journey.

"I agree to lead the six of you back-to-the-past to observe the creative process going on all over the universe and planet Earth." Dabar accepts the leadership. "Let us wait a bit before we begin our trek. Let us wait until November 22, 1.6 bya, to begin our journey. In the meantime, you can carry out your present mission to improve and coordinate all of the new functions and systems recently acquired by Prokye 1. You can also continue to make new reproductions of the latter with all the improvements passed on to offspring."

The twelve friends begin their new mission full of hope. The six members selected for the next journey are encouraged by hearing the stories of those who have already gone back to the past.

Chapter 44
Back to the Past: Dabar
Leads Six Atom Friends on a
Journey

"I am ready to go!" Hydrojohn shouts as Dabar becomes present to the crew of Prokye 1.

The time is November 22, 1.6 bya.

As Dabar promised, he comes to lead six of the atom crew member on a journey back to the past. He begins, "I am excited. I would like to begin by taking you back to the origins of the universe on the first moment of January 1. Let us begin."

"Egregious expectation fills my bedazzled being!" Photie anticipates an overwhelming experience.

With unbelievable speed, Dabar leads them back 13.4 billion light years to the moment just before the beginning of time and space. All is blackness, except for an intensely radiant realm.

"Dabar, what is that intensely radiant sphere beyond this awful darkness?" Nitrojoan is fascinated by its scintillating white-crimson glow.

"We, the Divine Beings, live in that timeless realm," Dabar says. "Be attentive! Momentarily, you will watch the beginning of the universe and space-time."

No sooner had he spoken than from this majestic realm came an immense explosion of bright reddish energy which creates, within a brief moment, a colossal bubble-womb.

"What is that gigantic energy force emanating from your mystical realm, Dabar?" Oxyjoy asks.

"That limitless energy is our ardent love radiated outward, driving back the blackness as Ruah births the universe," Dabar begins. "We, the Divine Beings, are sending our love energy forth to create beings to which we will be able to relate." He pauses and then continues. "Do you observe Ruah now hovering over the universe?"

"Yes, I see our ghostly godmother, Ruah." Hydrojohn is overawed at the explosive sight. "When I observed this gigantic explosion of crimson-red love energy, I expected to hear the roar of millions upon millions of thunderclaps. This rapturous love energy has been birthed forth to the rhythm of soft, gentle music building up to a crescendo of a joyous, harmonious melody. I hear this melody reverberating throughout this expanding universe."

"Wherever our ecstatic energy of love moves in our universe, it is signified and accompanied by this melodious symphony of sound." Dabar affirms Hydrojohn's experience. Then he beckons them to follow. "Let us enter inside of the reddish bubble-womb of the expanding universe." Dabar pauses. "In space-time, it is only a minute part of the first second of creation."

"Dabar, the temperature of the universe is tremendous," Hydrojohn comments. "The clash of these massless packages of love energy is creating quadrillions upon quadrillions of white and black particles. The density of the latter is immense. Ruah is stringing her love energy between them. I recall Ruah naming this force gravity."

Oxyjoy reports what she observes. "As you mentioned, Dabar, intoxicating, rapturous music is accompanying this outpouring of divine love energy."

"Those white and black particles are made up of various shapes of curved loops and strings. These vibrating loops and strings create this lovely music," Dabar says. "Notice! Gradually, the black particles are being destroyed; only the white ones remain."

Nitrojoan interrupts Dabar. She shouts, "Photie, I see you being created as the first yellow package of energy!" Nitrojoan shouts, "The Divine Beings are talking to you. You do not stop. You speed off into the expanding crimson bubble of the universe."

"Frightening fear caused me to flee," Photie says. "A miserable memory, but a source of peaceful pride with happy hindsight."

"Look! There is Quarkie with the Divine Beings. She looks very fearful." Oxyjoy sights her new friend. "She is joining two other quarks. They must be Quarko and Quarkoff." She is amazed by this past event. "When I arrive back home, I must tell the quark trio about being present to their birth."

"It's the end of the first second, my friends," Dabar announces. "The bright, crimson-red universe is expanding at a slower rate. The temperature is descending rapidly in intensity from the initial trillions of degrees. Now it is fifteen minutes into creation history. The temperature is in the tens of millions of degrees." Dabar continues his observations. "It's the third hour now. The temperature is about three thousand degrees. Can you see the bright crimson glow beginning to diminish to a soft sunset color?"

"The universe is now black," Oxydon comments. "Is this the end of the universe?"

"No, Oxydon," Dabar responds. "Notice, the quark trio have created the first white proton." Then Dabar asks, "What is happening with all of these new beings?"

"All the members of the quark family are joining into communities of three quarks to form white protons," Phosie answers. "Just before the universe went dark, I observed gigantic collisions between all of the particles and yellow packages of energy, the photons. The clashes were giving birth to amber electrons, white protons, silver neutrons, emerald neutrinos, and yellow photons. Mass was being converted to energy and energy into mass." Phosie pauses and muses. "Oh, one more remembrance. I noticed the amber electrons began being drawn by love energy to enter into a relationship with the white protons to form the first orange hydrogen atoms. To the rhythm of a passionate melody, they began to multiply in stupendous numbers as the force of this love drew them into this relationship."

"I saw white protons being drawn by an intense love force to a silver neutron to form the first helium nuclei." Oxydon recalls what he has just seen. "The melodious energy of love was drawing protons and neutrons to create immense numbers of the helium family."

"I see the multitudinous members of the hydrogen and helium families of atoms being drawn to form islands of hundreds of millions of gaseous clouds," Oxyjoy reports. "Ruah has been overarching all of these creative activities on the neophyte islands filled with hydrogen atoms and helium nuclei. Elohim called the islands of clouds galaxies."

"My dear atom friends, we are moving forward now to January 25, 1.0 billion years into creation history (byich)." Ruah is directing them to a new destination. "Let us enter into this special part of this huge, gaseous cloud nebula to proceed to its core." They arrive there. "Notice that the temperature is rising rapidly due to the spinning action of this part of the cloud and the pressure exerted by the hydrogen atom family on its core."

"Look, there is Quarkie, Quarko, and Quarkoff!" Nitrojoan shouts her recognition of their friends. "They are helium nuclei now. They are working to create. What will kind of being will it be?"

Suddenly, the core ignites into a golden, white-hot flame. Yellow-white radiation shoots through this gigantic gaseous sphere peopled by the hydrogen family.

"Quickly!" Dabar orders. "Let's move out beyond this spinning body." They move at tremendous speed to an observation point beyond this body.

"Look, this body is igniting with stupendous light." Oxydon describes his experience. "The blackness of the universe's bubble-womb is lighted by this new radiant sun. It is like our Sun."

"Yes, Oxydon and my friends, this new sun is the prototype of all of the hundreds of trillions of suns that will exist in all the billions of galaxies," Dabar explains.

"The quark trio, as part of the helium family, assisted by the hydrogen family, cocreated this first sun." Phosie is amazed to have been present to their cocreating.

"In this past time, I am going to transport you quickly forward to February 1, 1.3 byich," Dabar instructs. "What do you notice?"

"From January 25 to February 1, or 200 million years in time, the hydrogen and helium families have spun out billions of new stars," Phosie responds. "When I look around our universe, the billions of galaxies look like lighted communities inhabited by hundreds of billions of suns giving some light to the enormous blackness of space-time."

"The love energy of gravity unites all of the galaxies, its suns, and gaseous clouds of atoms," Nitrojoan reports. "Ruah is everywhere as creative directress."

"Dabar, what is this nearby strange, dark red sun?" Oxydon inquires.

"Let us enter the core of this vermilion sun," Dabar directs them. They travel to its core. "What do you observe now?"

"The furnace at its core seems to be dying out." Oxydon ventures to respond. "The bright yellow-white of this first sun is now a sputtering, reddish sphere."

"Hey!" Phosie shouts out her recognition. "There is the quark trio amidst this dying core. I think that the mountainous hydrogen gases in this huge, reddish sun are going to come crashing down on this core. You can feel the pressure growing."

"If the quark trio is here, this must be the same proto-sun we saw at its birth. How can that be?" Oxydon is confounded.

"My friends, your observations are correct. This is the proto-sun you saw at its birth." Dabar begins his explanation. "Some suns live for billions of years before getting old and dying. Some suns that are very large at their origin age rapidly. They run out of members of the hydrogen and helium families to fuel their furnaces. They begin to die. This red sun is close to death." He pauses and continues. "Let us travel outside of this gigantic red star and observe what happens." They move to an observation point some distance away from this dying sun.

"You were correct, Phosie; the exterior of this sun is crashing in on its core furnace." Oxydon affirms her observation.

"Wow! It is exploding!" Oxyjoy shouts.

"This is the first supernova, the prototype." Phosie is enthusiastic. She reminisces. "It brings back the memory of Ruah's prophecy of the 'provoking event' that brought about the origin of our Sun and solar system. It was a supernova that energized our Sun and brought into existence our planet Earth."

"What more do you observe?" Dabar asks.

"There are new families of atoms being created in the explosion of this red sun." Nitrojoan is surprised to see these new families.

"There is the quark trio. They are no longer helium, but are now a carbon atom," Phosie says. "I remember queasy Quarko remarking that he was a quick-change artist. That explains his remarks." She adds another comment. "I like journeying into the past. I am able to see all of the cocreative work being done by all of the atom families.

"You are correct." Dabar adds to their observations. "Supernovae create all the ninety-two atom families through the creative work of hydrogen and helium. The explosion of these red suns then seeds the universe with all of these ninety-two elements. Of all of the latter in our universe, hydrogen makes up 80 percent of all atoms, helium 19 percent, and the other ninety make up just 1 percent."

Dabar says to the sextet of friends, "We have completed this part of our journey back to the past. Let's move forward to September 10, 4.6 bya."

Chapter 45
Back to the Past: Dabar
and Friends Continue Their
Journey

"This is a familiar scene, Dabar." Phosie begins the dialogue as they reach the vicinity of a gigantic nebula on an outer arm of the Milky Way galaxy.

The time is September 10, 4.6 bya.

"That immense red sun is commencing to explode." Oxyjoy alerts her crew mates.

"The 'provoking event!'" Photie comments with serene calm. "Sensational supernova blasts precocious planet Earth into bountiful being!"

"Yes, precious Photie, this supernova began the process of creating the sun, your Sun, and all of the planets in your solar system." Dabar confirms Photie's identification of the event.

"Thus, how wonderful in our journey into the past to again be present to this magnificent event, Dabar." Nitrojoan is filled with gratitude. "Thank you, Dabar."

All of the six crew mates watch the birthing of the Sun and the beginnings of the four rock and five gas planets. Dabar leads them on a swift journey. As they are present to the solar system on September 25, 4.0 bya, they observe all of the nine planets taking form. They notice that the bombardment of astral bodies into the Earth is diminishing. All over the Earth, they see huge volcanoes belching out magma, water, and gaseous material. Thus, they note that the atmosphere of Earth is ripe with water vapor seeded with various gaseous molecules that are dark gray in color with some flecks of orange. To some degree, they block out the Sun's radiation. They pass by many small islands of volcanic-paved crust. They feel the Earth cooling. As a result, they see the waters in the air condensing and falling as torrents of rain upon the basins of the oceans and earthen crusts. The atom visitors are amazed that the massive amounts of water are gradually filling the far-flung, gigantic-sized oceans. They stand in awe of the constant downpour of water carrying with them many different chemical molecules, which constitutes a rich molecular soup.

Dabar moves them to October 5, 3.6 bya. During the past ten days, September 25 to October 5, or 400 million years, the six shipmates observe the oceans growing in volume due to the continuing torrential rains, amidst stupendous bolts of lightning and deafening thunder claps. The shores of the oceans appear to them as a dense, molecular soup. They visit the islands of crusted rock and travel to the bottom on the ocean floor near the multitude of vents and cracks where volcanoes belch forth their ocean flooring magma. Dabar and his six friends finally settle on a single area of the seashore.

"I see myself." Phosie is surprised. "We are working to co-create new beings. There is the quark trio, Nitro, Hydrojean, and Photie."

"The sonorous sound of majestic music appears to be impelling the energy of love at a deep level to assist in the birth of these new beings," Nitrojoan observes.

"Ruah hovers over this scene. On a physical, chemical, and molecular level, the atom families are creating these neophyte creations. At a much deeper level, Ruah is vivifying and enlivening these new living beings." Oxydon is transfixed on the scene.

"Phenomenal Prokye 1, the primal prokaryote bacterium, is birthed to luminous life." Photie alliterates this awesome birth scene.

"Do you observe the atom community now working on another project within Prokye 1?" Dabar asks.

Oxyjoy is quick to answer. "They are working to develop RNA nucleotide bases. I notice them now working to create DNA nucleotide bases to take over one of the dual functions that RNA carried out in the beginning."

"Did you ever realize in your wildest imagination that when all of you helped to cocreate prokaryote bacteria that you would help your creations to be able to later make reproductions of themselves?" Dabar quizzes the six.

"Never did I dream that our cocreative activity would be able to make duplicate copies of itself." Oxydon is amazed. "Not only that, but that they would have made, through our creativity and the loving energy of Ruah, continuous modifications to their design plans as time proceeded."

Dabar accompanies them further on their journey back to the past. They move forward to October 22, 2.9 bya.

"As we traverse along the shores of the oceans that surround the islands of magma crust, what do you observe?" Dabar is testing them.

"The shores are rich with a wide diversity of prokaryote bacteria," Hydrojohn answers. "From Prokye 1 to this expanding variety of prokaryotes, the DNA and RNA bases have demonstrated their ability to modify their design plans with the assistance of our co-creative work."

"The numbers and diversity of prokaryote bacteria are staggering, Dabar," Phosie notes. "The watery environment outside of the bacteria community and the semi-aqueous fluids within the walls of each individual prokaryote contains a rich variety of atoms, molecules, amino acids, nucleotide bases, and proteins that can be used in our co-creative work."

"What are those layers of mats?" Nitrojoan asks Dabar.

"You probably notice that prokaryotes must dig into the mud or stay away from the surface of the water so as not to be annihilated by the Sun's radiation." Dabar begins his explanation. "So the inventive atom families have helped the prokaryote beings to be able to live above the water in these mats at the edges or in the waters of the seashore. We have named these mats stromatolites. Large communities of diverse bacterial beings live there," he concludes.

"Let us move forward to October 30, 2.5 bya." Dabar leads them forward. "Let's check out the atmosphere."

"The grayish orange skies have been fading over recent days." Nitrojoan makes the observation. "With the oxygen atom families beginning to seed the air with gaseous oxygen and the nitrogen atom family releasing its gas into the atmosphere, the sky is becoming at times a bright blue color during daytime on Earth."

"The torrential rains of early Earth have given way to periodic rains that refresh the ocean waters and wash the crusts of the Earth of salt and other minerals and carry them by the rivers back to the oceans," Oxydon notes. "This seems to be a part of the recycling program being worked out by the atom families."

"A wonderful strategy." Dabar is pleasantly surprised by his atom friends' observations. "What do you observe about the crusts of the Earth's surface?"

"The crusts are still islands of rock that continue to grow due to volcanic activity, which paves more of the surface." Phosie is quick to respond. "I would guess that the oceans make up 80 percent of the surface area and the crusts only 20 percent."

"Dabar, I notice that our creation of photosynthesis in Prokye 4 has caused a proliferation of prokaryotes that possess this ability to harvest the ultraviolet rays of the Sun as a source of energy and food and release, as waste, the oxygen gas into the atmosphere." With justifiable pride, Hydrojohn observes the activities of the bacterial beings.

"An important observation, Hydrojohn. I believe you, Oxyjoy, wondered what the waste product of photosynthesis, oxygen, would do for the Earth. Now you know. It is being utilized by the atmosphere. Diverse prokaryote bacteria are learning to live in this atmosphere." Dabar pauses. He changes the subject. "My dear friends, let's journey forward to November 12, 2.0 bya." Dabar again leads them forward.

"My oxygen family is working in the upper atmosphere, Dabar. What are they doing?" Oxyjoy inquires.

"The best way to find out is to talk to them," Dabar advises. They move in the direction of the atoms' activity to talk to their leader.

Oxyjoy begins the conversation. "Greetings, dear members of my oxygen family. My name is Oxyjoy. With five of my friends from various atom families, Dabar is taking us on a journey to see all the cocreative work going on in the universe and our solar system."

"Good to meet family members, friend." The leader returns their greetings.

Oxydon is anxious to meet members of his family also. He introduces himself. "I am Oxydon, a relative." He turns to Oxyjoy.

"Could I ask a question?" She agrees. "What cocreative effort are you carrying out?"

"We have noticed that bacterial life on the surface of the waters of the Earth is threatened by the intense rays of our Sun," the oxygen atom explains. "We decided to try to protect the bacterial family members by building a filter to cut down the intensity of our Sun's radiation. This filter is made up of three atoms of our family members. We believe that it will effectively accomplish our purpose to protect the bacterial community."

"Have the Divine Beings given your cocreation a name?" Oxyjoy asks.

"Ruah called our filter ozone," the oxygen atom recalls.

"Thank you, friend." Oxyjoy is grateful. "So our oxygen family is also working in the atmosphere. On our journey, we have noticed that the prokaryote community is modifying its design plans so as to be able to live in this growing, oxygenated atmosphere. So you, my friends, are contributing to their safety by your protective shield. You can be very proud."

"One other observation, Dabar," Hydrojohn adds. "I notice that the crust of the Earth now represents about 30 percent of its surface."

"By the way, now is the time to move forward to November 22, 1.6 bya. We will have a chance to check on your observation then. Off we go!" Dabar directs.

Dabar and his six atom friends arrive on a portion of the seashores. Prokaryotes continue to multiply and diversify in stupendous numbers.

"We can now journey into the magma to ask some questions about why the crusted surfaces are growing in size." Dabar reminds them of their concern.

Within the magma, they greet some silicon and aluminum atoms hard at work at their job of recycling. Hydrojohn greets them. "Friends, my name is Hydrojohn. I am accompanied by Dabar and five of my friends." Greetings finished, he asks, "I notice that the surface crusts of the Earth are growing larger. What is happening?"

"The vast number of volcanoes is helping to recycle the materials of our Earth," the silicon atom begins. "The liquid magma pours out of the cracks in the crust and paves both the Earth's surface and the ocean floors. There is a new plan being devised for our recycling."

"Please tell us the plan." Phosie is anxious to hear the new development.

"We want to be able to recycle the ocean floors and the Earth's crust," he notes. "We have labored to build the crusted surfaces into plates or platforms. They float on the red-hot magma, as do the floors of the oceans. Where ocean floors meet the crusted surfaces, one of these surfaces will be pushed down while the other slides over it. Those surfaces, forced downward into the fiery red magma, return to a molten state. The very heavy atoms move downward in the magma, the lighter atoms toward the top."

"How does this aid the recycling effort?" Hydrojohn asks him.

"The melted recycled atoms stay in the magma." He continues. "Later, they are belched forth from the volcanoes in the bottom of the oceans and the crusts of the surface. Not only are mineral atoms returned to Earth, but also salts and water that have been recycled in the magma."

"That is very interesting," Oxydon comments.

"Friends, we do not think that this is the end of our experimentation at recycling." He points to the possibility of further creative innovations. "An important process of recycling has been put in place."

"Dear friend, thank you for your guided tour underneath the crusts of the Earth. We are very impressed with the cocreative activities of you and all the atom families, not only in our solar system, but also the whole universe." Oxydon reflects on this present journey. "The energy of love and Ruah's encouragement is moving the cocreative process forward." He turns to his host. "Thank you for your hospitality. For all of us, I bid you farewell." The six atom characters depart with Dabar.

"Where will this all end, Dabar?" Nitrojoan is concerned. "We have labored for 13.4 billion years. I wonder when we can rest on our laurels."

"We have always encouraged you to stay the course," Dabar comments. "I feel deeply that your cocreative work holds many surprises and wonders, my friends." After a pause, he adds, "I believe it is time to return to the present."

"I am ready to rejoin our friends." Oxyjoy is anxious. "I cannot wait to tell our friends we saw them in action during our trip to the past."

"Thank you, Dabar, on behalf of all of us, for taking us on this special journey." Nitrojoan initiates the words of gratitude. "Dame Destiny and Radiant Ruah will lead us to continuous creativity!" Photie pronounces his prophetic proclamation.

Chapter 46
Divine Beings Dialogue with Their Atom Friends: Where Do We Go from Here?

"I observed you, Quarkie, Quarkoff, and Quarko, being created, and later, working to cocreate Prokye 1, during my journey back to the past," Phosie says with excitement. She looks to their response.

"I know that you observed me, Phosie, outshining my quark mates in creativity." Quarko is up to his old tricks. He trumpets his supposed creative superiority over his companions. Phosie does not expect this response.

"You are back to your old habits, Quarko." Elohim comes on to the scene to comment. He is accompanied by Sophia, Dabar, and Ruah. "Let me emphasize to you, old friend, that you participated with all of the atom families in all the cocreative activity that brought about the development of the universe and your planet Earth." He gazes deeply into Quarko's being and adds, "Ruah has encouraged and energized your creative work through the love energy that has its source in us, the Divine Beings. I remind you again: it was your ingenuity and persevering effort that were decisive in cocreating the universe and planet Earth. Congratulations for your wonderful, creative efforts."

Quarko is subdued and speechless at this moment. Elohim speaks the truth in love to him. The other eleven atom characters are angry at Quarko because of his haughtiness but feel that Elohim's reprimand is a sufficient remedy.

"We have been discussing our journeys back to the past." Quarkie summarizes their recent dialogue. "We are all amazed at observing, up close, the vast web of creation. Gravity that originates out of your love energy, oh Divine Beings, binds all beings in our universe into an intimate unity."

"It is almost a paradox," Hydrojean comments. "In the macrocosm of the universe, the billions of galaxies, with all of their hundreds of billions of suns, are united by this weak, intimate love energy to all of us who live in the microcosm of the Earth. It is unbelievable to contemplate."

Oxydon takes up the dialogue. "As we traveled the universe, we became aware of your love energy. The strong force, the weak force, and electromagnetism, which are alive and working within all atoms and other created beings, also find their origin in the divine love energy that bursts forth from you, the Divine Beings, at the time of the birthing of our universe home. Your power of love energizes at a profound level all of our creative work."

"I am especially aware of a love principle that has guided all of our cocreation, a loving spirit of cooperation." Oxyjoy speaks from experience. "On my journey, when I spoke to other members of our oxygen family, they exhibited this same cooperation."

"I had the same experience of loving cooperation toward a common goal when I met other members of my phosphorus family during my travels." Phoson agrees with her. "What an amazing diversity of beings emanate from this cooperative, creative effort." He is exultant.

"The community of prokaryote bacteria has gone through immense diversification," Hydrojohn remarks. He goes on to describe a recent back-to-the-past experience. "I saw one community gathered together near a seashore. This slimy, green community contained a myriad of diverse bacteria. Through an alliance, they built various sizes of mated structures above the water. There they built pathways for water and nutrients to pass by the homes of diverse bacteria to help feed all the members of the group. The waste products of each community had another pathway. What was waste to one group was food for another. The most honored bacteria were the ones that contributed most to the cooperative effort." Hydrojohn was emphatic in making his point.

Oxyjoy gives another example. "During our first visit to a larger prokaryote, we discovered an openness to our visit and our sharing with them of our cocreation, the process of photosynthesis. We only had time to transfer our DNA template containing the process's design through an RNA messenger to our host's DNA template. Then the crisis of insufficient nutrients for both of us arose. We had to abort our mission." She concludes with this observation, "Near the end of our recent journey, I saw quadrillions of bacteria that were using our creation to receive the energy from our Sun's photonic radiation and using this energy to convert various molecules into food in order to sustain the prokaryote community."

"At the same time, Oxyjoy, this process of photosynthesis is releasing two atoms of your oxygen family as waste," Hydrojohn observes. "This waste oxygen is being utilized by some of the diverse bacterial families as energy."

"So rather than oxygen gas in the atmosphere being an enemy of prokaryote bacteria, the latter have been able to adapt in order to use this gaseous element as an ally." Nitro begins to glow with pride. "Our nitrogen family is also contributing to the atmosphere of our planet Earth." Nitro ponders briefly and says, "What wonders the cooperative efforts of the ninety-two atom families have accomplished!"

Sophia is enthralled by the frank and lucid observations of the atom friends. She compliments them. "Your powers of observation are outstanding, dear friends. You have captured the spirit and the reality of the cooperative attitude of the ninety-two atom families and the superabundant photon family that have cocreated the beginnings of our magnificent universe."

"Oh thank you, thank you, sovereign Sophia!" Photie exudes gratitude and love. "You have remembered our prodigious photon family and our august accomplishments."

"Yes, Photie, we have been negligent in commending your photon family for its 'awesome accomplishments.'" Nitrojoan is apologetic.

"The oversight has been corrected by Sophia, dear friends," Dabar says. Then he abruptly changes the subject. "You are all presently residing in Prokye 1. Have you made any new adjustments to the latter now that you have assimilated all of the new creations brought by the crew mates of Prokye 2, Prokye 3, and Prokye 4?"

Quarkoff begins a response. "The nucleus of Prokye 1 has been strengthened through the development of an improved membrane wall around the DNA helix. As you know, the mitochondria and photosynthesis receptor units have retained their DNA templates, but they have transferred copies of their templates to the Prokye 1's DNA helix."

"Phoson, Nitro, and I have worked to further perfect our spindle apparatus and kinetochores." Hydrojean recounts their efforts. "I still intuit that someday this spindle apparatus will help to further the cocreative process. We will just have to wait and see."

317

Quarko has been very subdued and quiet. He seems to be serious as he says, "Since it was through all of our efforts that we have enhanced the abilities of our friend, Prokye 1, I believe that we have created a whole new being." He hazards a further opinion. All the atom friends and Divine Beings are fixated on him. What will he propose? "There are no prokaryote bacteria that possess all of the functions and abilities that Prokye 1 has. Some of them have certain functions, but none of the diverse prokaryotes are like prolific Prokye 1." Quarko is coming to a conclusion. "I believe that Prokye 1 is a new being and should be given a new name!"

The eleven crew mates stand in awe at this proposal. They have not come to the same conclusion as Quarko. Begrudgingly, they admire the boldness of his plan.

"Your proposal has much merit." Ruah is the first to respond.

"A brilliant deduction, phenomenal Quarko." Elohim is lavish in his praise. "Yes, Prokye 1 is really a completely new type of being because it has membrane walls around its organelles, like the nucleus containing the DNA helix, the mitochondria, photosynthesis platforms, and receptors." There is an interlude, during which Elohim muses. Then he states with authority, "I will rename Prokye 1. It is now to be called Eukaryote 1, or a more familiar, homey name, Eukarye 1. No longer will Prokye 1 be identified as a bacterium. It assumes a new identity as a eukaryote cell and new name, Eukarye 1. I intuit that it will be forerunner of a diversity of new beings."

All direct their attentions toward Prokye 1. They wonder how it will receive this renaming. The taciturn Prokye 1 is accepting of this dramatic turn of events.

"Prokye 1, now Eukarye 1, you have been open and obedient to the cocreative process." Phoson is complimentary of its humble attitude. "Thank you for your spirit of willing acquiescence." It silently nods its acceptance.

"Where do we go from here?" Sophia proposes a question.

There is a pause. The twelve atom friends are in deep contemplation. The Divine Beings allow this time of silence. Finally, there is a stirring. Quarkie readies herself to answer Sophia's question.

"We are a web of creation. Gravity unites us in love. You, Ruah, continue to be our energizer. We can depend upon you to energize our cocreative work. You, Elohim, Sophia, and Dabar, are present to encourage and support us, ever since we were birthed into existence through your wish to share your love with other beings. These are the givens about our universe and its creative activity. To move the universe and planet Earth into the future time, we need to continue to be aware of the above realities of our existence."

"I look at the vast oceans and see bacterial life existing only near the shores of the oceans, seas, ponds, or rivers, and in the deep ocean near the volcanic vents." Nitrojoan contributes her ideas. "Is it possible that we could diversify eukaryote cells so that they could produce beings that could inhabit all of the areas of this enormous ocean?"

"Are there enough nutrients and molecules in the open oceans to be able to nourish such beings if we cocreated them?" Oxydon wonders.

"Can the offspring of Eukarye 1 grow larger?" Hydrojohn muses. "That would be a great leap forward. How would we be able to increase the size of Eukarye 1 enough to be able to meet the challenge of traveling the vast distances of the ocean?"

"In the past, prokaryote bacteria have produced one offspring at a time." Nitro poses a new challenge. "Would it be possible to join two eukaryote cells together? What about three or four or more? How would we do this?" He foresees a future dilemma.

"Possibly the further development of our spindle assembly and kinetochore could answer your question, Nitro," Nitrojoan proposes.

"You have named a possible future mission for Eukarye 1." Dabar affirms her. Nitrojoan beams with pride.

"The crusted islands of rock surface above the oceans and rivers are bare of bacterial life, except close to the seashores," Oxyjoy observes. "Could Nitro's multicellular beings, if they are possible to be created, be able to inhabit this harsh environment?"

"You have made some very valuable, interesting proposals, my dear friends." Sophia is pleased by the several proposals for future missions. "The future development of planet Earth, the solar system, and the gigantic universe depends upon the cocreative effort of you and all the other ninety-two atom families. Ruah will continue to be your faithful directress and support."

"Congenial cooperation and caring companionship must continue to enliven our magical mission of capricious creativity," Photie opines with calmness.

"You recently made another prophecy, noble Photie," Dabar recalls. He continues. "You said that Dame Destiny and Radiant Ruah will lead to 'continuous creativity.'" He adds, "Dear Photie! You have added an important addition to your prophecy:'congenial cooperation will lead to capricious creativity.'"

Epilogue

"Eukarye 1 is a cell and not a bacterium." Quarkoff seeks to adjust to the new reality. He turns to Quarko. "Much as I hate to admit it, you, Quarko, made a brilliant deduction. I was shocked when Ruah and Elohim confirmed your conclusion."

"I accept your compliment with humility, mate Quarkoff." Quarko seems humble for once. Faced with his seeming sincerity, his atom friends are in shock.

The time is still November 23, 1.5 bya. The twelve crew mates of Eukarye 1 are in deep conversation about the future. As they contemplate this future together, a spirit of pessimism and anxiety overcomes and clouds their former optimism. The missions of the future seem more daunting and overwhelming.

"Eukarye 1 offers new possibilities for furthering the diversification of new beings through our cocreative efforts." Phosie is encouraging.

"I have been enthusiastic about our creativity." Quarkie offers a note of gloom. "I am getting tired of all of this frenzied activity. I would just like us to stop and rest on our laurels." She seems to be returning to her old fearful ways.

"I feel the same way." Hydrojohn agrees.

"Our existence becomes harder and harder as new demands are placed upon us." Phoson affirms Quarkie's feeling of exhaustion. "Maybe we should approach Ruah about our dissatisfaction and deep concerns about the future."

It would appear that a majority of the crew mates are joining the rebellion against further cocreative work. Having cocreated Eukarye 1, they would like to rest and just enjoy the present state of the planet Earth and the universe.

Photie listens politely to their anxieties and concerns. "Felicitous, faithful friends, potent perseverance calls you to fastidious fidelity." Faithful Photie confronts their flagging spirits with gentleness.

"Bravo, dear friend Photie." Quarko takes up the latter's challenge to the rebellious group. There is a sincerity in his tone and authenticity to his words that has been so lacking in the past. He proceeds. "Now is not the time to stop our cocreative work. The future looms with great possibilities of dynamic creativity. I believe we are not too distant from special breakthroughs that will lead to unimaginable beings to populate planet Earth." He urges them to renewed constancy and greater efforts in pursuing their mission. "Courage, friends! Or as Photie says so eloquently, 'potent perseverance' and 'fastidious fidelity' to our mission! 'Congenial cooperation and caring companionship' is needed at this crucial time in the Earth's history!"

The rebellious atom friends listen carefully to Photie and Quarko. They are not swayed by their sincere urgings. I sense that this flagging spirit is gripping many of the members of the ninety-two atom families all over the universe. Is creativity on planet Earth about to come to an abrupt end? Will creativity all over the universe slow or cease?

Afterword
The Future of Creation

Some of you might wonder why Book I ends so abruptly with the creation and evolution of the eukaryote cell. This cell will be the foundation for all future flora and fauna, including human beings, who will inhabit planet Earth over the next 1.5 by, or November 15 through December 31. All flora and fauna will be constructed of eukaryote cells. This period will be "The Age of the Eukaryote Cell." At the same time, the tectonic plates upon which the dry land masses and seafloors are constructed continue their constant movement, changing the locations of continents and seas. Sea levels continue to rise and fall periodically. Climate is in a continuous flux. The oxygen and nitrogen levels are not yet able to sustain eukaryote multicellular life 1.5 bya.

Utilizing the literary genre of midrash, Book II will continue my original vision of how the universe was birthed into existence through an act of ecstatic love on the part of the Divine Beings at the first moment of space-time. My vision will continue to recount the adventures of Quarkie, Photie, and friends. The latter continue to create new beings energized, but not programmed by the Divine Beings. This allows for chance, contingency, natural selection, genetic copy errors, and manipulation by a tool kit of genes to be involved in their creativity. During the last 1.5 by covered by Book II, the Divine Beings will be in intimate, constant contact with the atom characters. Since the latter live in the microcosm of the universe (you would need an electron microscope to view their activities), the Divine Beings frequently take them on journeys back to the past to view the macrocosm of the universe and to come in contact and talk to other atom families carrying out other creative, evolutionary missions.

You may also wonder how my atom characters can turn up at just the right time to continue their creative activities. Remember that atoms continue to exist even though the living being in which they live dies. Such discarded atoms are assimilated by other living creatures and they become part of that being. In addition, the magic genre of midrash allows me to take liberties in telling the scientific story of creation without compromising the facts, artifacts, and conclusions of modern science.

In the beginning of Book II, the atom friends decide to keep their teams together. Team One is composed of Hydrojean, Nitro, and Phoson; Team Two of Phosie, Oxyjoy, and Hydrojohn; Team Three of Photie, Oxydon, and Nitrojoan; and Team Four of Quarkie, Quarkoff, and Quarko.

Ruah takes the atom crew mates on a journey to survey the present state of evolution on planet Earth, the climate, tectonic plates, bacterial and eukaryote communities, atom family members at work in the molten mantle underneath the plates of the Earth, on dry land, and in the atmosphere.

Team Three evolves new varieties of algae about 1.2 bya, or December 1. Team Two, working within eukaryote protists, creates a new process for the reproduction of progeny by union of two protists 1 bya, or December 5. Ruah names this process meiosis. Then, after suffering the frustration of many trials, failures, and some interpersonal conflicts over a period of 1 billion years, November 15 to December 15, Teams One, Three, and Four finally master the process of creating eukaryotic multicellular life, such as sponges and jellyfish. All of this creativity takes place in the waters of the seas, rivers, and ponds, but not on the barren soil of the Earth. Through their creative efforts, they begin to develop a tool kit of genes, such as hox genes and improved genetic-on-off-switches, which will help them in their future creativity. With the help of this tool kit, they create the two membrane flatworms, and then three membrane roundworms with bilateral symmetry, primitive eyes and mouth, the beginning of a brain and neural system, rudimentary heart, and vascular, breathing, digestive, and waste systems.

About 600 mya, or December 15, the atom characters, along with numerous atom friends, suffer through a period of an Earth-wide cold spell and an ice-covered planet, called "snowball Earth." The prokaryote, protists, and multicellular communities suffer some extinction of species. Creativity comes to a standstill. Will this catastrophe halt future evolution? Ruah takes Teams One and Four and they travel back to the past to view the destruction of species. They are devastated and depressed to observe the extinction of some of the life forms they labored to create more than 1 bya. They, too, wonder about the future of earthly creativity. The Divine Beings console and encourage them to resolutely continue their creative efforts. Ruah promises to power their efforts with her divine energy of love.

After much effort and cooperation from 600 mya to 540 mya, or December 15 to 17, all the teams are working diligently and steadily, supported by the Divine Beings' love energy, within this watery environment. Finally, they achieve success. Their creativity leads to a strong surge of multicellular creativity and evolution, called the Cambrian Explosion. Team Two creates and evolves trilobites, scorpions, mollusks, and later, Team One creates hemochordates or proto-fishlike creatures, and Team Four jawless, then jawed fish with a skeletal system. The tool-kit genes produce more sophisticated eyes, mouth, teeth, brain, heart, circulatory and neural systems, in addition to digestive and breathing systems, in all of the new species of sea life. At the same time, other atom families evolve plant life, which proliferates in the seas, rivers, and ponds during the Cambrian Epoch (545–495 mya). In the waters of Earth, Team Three evolve the first multicellular plant life, red algae, near the end of the Cambrian Epoch. At the same time, they gradually evolve the ability of green algae to survive in the atmosphere. The land masses are still devoid of any life except for some colonies of single-celled prokaryote bacteria.

The Ordovician Epoch (495–438 mya) begins. About the middle part of this epoch, green algae is able to go ashore, survive, and carpet the soil.

It is at this same time that the Earth suffers a second devastating catastrophe, which decimates several communities of marine life, including trilobites, which go extinct. Dabar takes Teams One and Two back to the past to view the Ordovician Age, its creative wonders, and the devastation caused by this second large extinction. The team members are very depressed by the loss of so many marine species, as are the atom community members that they meet on their journey. Dabar encourages them to continue their efforts despite this serious setback. The teammates are skeptical that there is a future for their creative activities.

The atom families take heart again. It is the beginning of the Silurian Epoch (438–417 mya). With the loss of the trilobites, after 100 my, they labor to improve the chordate (fish) species. Team One uses some of the tool-kit genes to evolve a more advanced skeletal system, jaws, and backbone, to strengthen them.

Algae stranded on dry land by receding waters adapt to these new conditions and develop new species to further populate the land areas. Oxygen and nitrogen atom communities stabilize in the atmosphere around the Earth, while the oxygen community further develops a layer of ozone in the upper atmosphere to protect living beings on the Earth from the sun's deadly ultraviolet rays.

Dabar accompanies Team Three on a journey. The temperature, which cooled near the end of the Ordovician Epoch (438 mya, or December 18), is gradually warming toward the end of the Silurian Epoch (417 mya, or December 17). At the beginning of this epoch, all species of biological life, except algae and some prokaryote colonies, live in the waters of Earth. They observe Team One creating new species of algae. They observe the latter move onto the soil of the Earth near the shorelines and survive. Most of the continental plates still reside in the southern hemisphere of the Earth.

On their return, Team Three decides to continue the efforts of Team One, who evolved algae that later moves onto land. They work to create a eukaryote cell that will help plant life to stand upright. Through disappointments, conflicts, and community effort, they evolve lignin, a wood cell, that assists plant life to stand upright. They also develop a vascular system to allow the moisture to move upward from the roots to bring water to the trunk, branches, and leaves, which they gradually evolve.

At the beginning of the Devonian Epoch (417–362 mya), Elohim and Sophia conduct Teams Two and Four on a journey from the end of the Silurian Epoch and the middle of the Devonian Epoch (385 mya, or December 21). The climate is mostly warm and dry, with cool temperatures near the south pole. Pangea, the coming together of the main crustal plates, begins to develop during the middle of the Devonin Epoch. Carbon dioxide levels in the air are high. This condition causes plant life to proliferate on the land masses, especially close to the seas and ponds. The upper atmosphere ozone continues to protect species from ultraviolet rays.

Team Four, along with many other atom communities, labors diligently, amidst conflicts between Quarkie and Quarko, to create and evolve the coalacanth with elementary lungs and an improved breathing system. Again, with the clever use of the tool-kit genes, they then evolve from fish an amphibian that evolves elementary legs from its fins. They propagate by laying eggs near the seashores.

This helps Team Four to think of new future creations and the possibility of movement of these amphibians to land.

Upon their return from their journey, Team Two creates and evolves insects with wings in the watery areas. The insects soon move on land, which is now providing sources of food.

In the latter part of the Devonian Epoch, there is a minor extinction. This catastrophe effects most marine life but not terrestrial flora. Reefs of coral and invertebrate fish life suffer great losses.

After the Devonian Catastrophe, Elohim and Sophia take Teams Two, Three, and Four on a trip back to the past. It is also the beginning of the Carboniferous Epoch (362–290 mya). They observe possible niches for new species opened up by this extinction. They observe the activity of the continental plates, the climate, and the atmosphere of planet Earth at this point in its history. By the middle of this epoch, extensive rainforests cover the tropical regions of the newly forming super-continent of Pangea. The south pole is developing an ice cap. The team members meet the atom families living in amphibians, who are developing a plan to evolve a harder shell for their eggs to enhance survival of progeny and a more efficient lung for living on the land.

By the middle of the Carboniferous Epoch (362–290 mya), Team Three evolves some plant species into large ferns and trees of various kinds. They continue to develop more sophisticated spores and gametes for reproduction of tree species. In some tree species, they evolve a hard embryo seed that contains its own food source for its reproduction cycle. The seed, which Elohim calls a gymnosperm, is a vast improvement over the soft-shell seed.

Team Four evolve on land the insect family, including beetles, and shell creations, like snails.

The Permian Epoch (290–240 mya) begins and moves forward. During the middle part of this epoch, Dabar takes Teams One, Three, and Four on a journey back to the past. The large continents are coming together to form the super-continent of Pangea, one part north, the other south, with a small sea in between the two. Giant, swampy forests grow on the tropical land masses as a suitable environment for the amphibian and insect communities. They meet other atom communities who are evolving larger, more diverse species of amphibians through the manipulation of the tool-kit genes. They discuss these new creations with them. Because the continents are joined, they notice that many species are able to move to new areas of Pangea and most of the continental areas of the Earth. They meet another atom community that has created a new species of fauna, a reptile, from the amphibian species. Being consumed with their own creative activity, they are surprised by the rapid creativity and evolution proceeding in all areas of planet Earth.

Near the end of the Permian Epoch (248 mya, or December 26), there is a rapid global warming due to the dissipation of the ice in the south polar region, which contributes mightily to a "hot house" Earth and a most devastating catastrophe, the Permian Extinction. Ninety-five percent of all the species on the land and in the sea are wiped out.

Elohim and Sophia escort Teams One, Two, Three, and Four on a journey to observe this momentous catastrophe. Their spirits sink when they assess the huge dimension of this extinction—the loss of 95 percent of all sea and land species. Again, like team members viewing other extinctions, they despair that they can continue their creative efforts. Why struggle through conflicts, obstacles, dangers, and strenuous effort to create an evolving flora and fauna, then watch it wiped out by devastating catastrophes like the Permian Extinction? Elohim and Sophia muse deeply as to whether they will be able to resurrect the demoralized atom communities to continue their creative efforts. But again, they encourage and console them to continue their evolutionary efforts. Reluctantly, Team Three, sparked by Photie, and later Teams One, Two, and Four acquiesce to their encouragement. Ruah is nearby, energizing them through her divine love energy.

The Triassic Epoch (248–206 mya) arises out of the devastation of the Permian Extinction. Reenergized, Teams Four and One evolve from the reptile family a new species of fauna called generically, dinosaurs. Team Four creates a giant, long-necked, vegetarian browsing dinosaur, which Sophia calls a sauropod. Team One evolves a smaller, ferocious, speedy dinosaur, a meat eater, which Sophia calls oviraptor. When they meet, Team Four and Team One are conflicted. Team Four asks why Team One creates a more efficient meat-eating species that will attack and live off of killing other vegetative species. This seems counterproductive to their creative mission. They had faced this problem in the Cambrian Epoch, when the trilobites preyed and ate other species of marine fauna. Is this part of the nature of creativity and evolution, they muse?

Around the same time, Team Three creates, by tool-kit gene manipulation, another form of fauna, a mammal, from the smaller reptile community. Once evolved, they begin to proliferate and diversify. This epoch also saw the continuous development of new species of plants and trees. Pines and ferns predominate in most of the land masses of Earth. The mammal and insect communities continue to evolve new species. The super-continent allows the large and small fauna to migrate to all of its areas.

With the beginning of the Jurassic Epoch (205–142 mya), Dabar takes Teams One, Two, and Three on a trip back to the past. They are surprised by the diversification and growth of the dinosaur community. Plant and tree life also continues their growth and diversification. They continue to provide the increasing dinosaur communities with food and shelter. Pangea is beginning to break apart. The equatorial land masses are arid, but the northern and southern hemispheres are warm and semi-tropical. South and north poles are becoming cool and temperate. Moisture-bearing winds are lush and verdant. Team members meet atom communities that are helping some of the dinosaur species to develop protection against the meat-eating dinosaurs. In the early Jurassic Epoch, another atom community evolves from smaller oviraptors a more ferocious dinosaur, which Dabar calls a tyrannosaurus. Since Team One was involved in the creation of oviraptor, they discuss this event with members of that community. The issue of dinosaur-eating dinosaur again arises. Small mammals evolve and diversify. They live in holes, trees, and caves. The insect community, especially the flying insects, also diversify more and more. Team Two, which evolved insects with wings during the Carboniferous Epoch, is especially interested in a development brought about by another atom community, a flying dinosaur, which Dabar calls a pterosaur. They share their creative stories with each other.

The large continental tectonic plates, which will later be North and South America, gradually separate farther and farther from the European and African plates. This is the time of the Creataceous Epoch (142–65 mya). On land, the dinosaurs roam on most of the continents of the Earth. While the dinosaurs further diversify and grow larger, some with bodily armor, the atom families evolve huge marine reptiles, one, the pliosaur, the tiger of the sea. Forests, rain forests, and swampy areas proliferate to be the environment and food source for all of these amphibians and large, dominant dinosaurs.

About noon on December 28, Team One creates and evolves from plant life, flowers, which begin to add a kaleidoscope of colors along with perfume scents to all the landscapes, attracting flying insects, which feed and propagate the flowers' seeds. This leads to a partnership between the insects and the flowering plants. This is an important step in evolution as it leads a little later to the evolution of vegetable plants. Toward the middle of the

Createceous Epoch (105 mya, or midnight December 28), Team Three evolve from small mammals a bird species that is a smaller version of flying dinosaurs. Near the end of this epoch, some of the dinosaur species begin to die out.

This epoch ends with a mountainous asteroid invading the solar system and crashing into an area just off the Yucatan coast in present-day Mexico. The explosion sends a gigantic amount of debris, soil, rock, water, and vegetation into the atmosphere. It forms a mushroom cloud of this material, which is carried around planet Earth for a long period of time. Trees, plants, and vegetation, derived of sunlight for a significant period of time, are stunted or die. The dinosaurs' food supply is seriously compromised, leading to the demise of the dinosaur community.

Elohim, Sophia, Dabar, and Ruah journey with Teams One, Two, Three and Four back to the past to view the massive destruction, later to be called the K/T Extinction. The team members are overwhelmed with frustration and deeply depressed at viewing this vast, earthly cemetery for all of the dinosaurs and large reptiles living on Earth. They confront the Divine Beings in anger. All seems lost. They are disillusioned and despairing. They point out the futility of continuing their creative activities, since periodically the Earth and its many species of flora and fauna are subjected to large extinctions through natural catastrophes. The Divine Beings observe that after each extinction, new and improved flora and fauna are created and evolve. The atom team members question the final goal of their creative activity. They are tired and want to retire from their creative mission. Photie rallies his colleagues to move forward to new and better creative activity with these words: "Friends, potent perseverance and fastidious fidelity to our mission is necessary." His encouragement leads them to discuss the possibilities for the future, using the living remnants of this extinction as the vehicle for new and possibly more sophisticated creativity. They commence some preliminary planning.

It is the beginning of the Paleocene Epoch (65–55 mya). On the continent of Africa, Team Four lives within one of the small surviving mammals about the size of a modern squirrel, that Dabar names a prosimian. These mammals propagate by means of an embryo that grows within a placenta in its body, and in due time, delivers offspring outside of its body. Team Four also works with the other atom families, through the manipulation of the tool-kit genes, to enlarge the brain and move the spinal cord connection to the brain more to the center of its head. They are conflicted as to increasing the size of this mammal but decide to stay with smallness.

Near the end of the Paleocene Epoch (56 mya, or the early afternoon of December 30), Dabar takes Teams Three and Four back to the past to observe the entire Paleocene Epoch. They notice that the North American tectonic plates are still attached to the European plates and South American plates are still disconnected from the North American plates. Temperatures rise after the K/T Extinction and, worldwide, are more moist and humid. Grasslands are covering many parts of the soil toward the end of this epoch. Forests are plentiful, vegetation proliferate, and flowers provide food for insects who, in turn, pollinate the flowers. They observe their prosimion creation and rejoice that it is prospering in the forests, where the climate is warm and moist. Teams Three and Four meet and talk with the atom communities who are evolving new species of plants, flowers, trees, and vegetation. They observe the creativity of still other atom communities who are evolving changes and diversification in the fauna community. They observe much smaller reptiles, amphibians, birds, and insects who survived the great extinction. They meet a community of atoms who are evolving a member of the marsupial mammal family and discuss their creative activities.

As the Eocene Epoch (54–34 mya) begins on the continent of Africa, Team Three live within a prosimian. They are debating how to evolve and develop this creature. Having noticed the forests and vegetation on one of their journeys with the Divine Beings, they decide that developing its hands and feet would benefit them in climbing trees and manipulating objects. In addition, they evolve a little bigger brain and develop a place near the middle bottom of the brain where the spinal cord enters. This is to help them in sitting and hopping in a more upright position. They use the tool-kit genes to evolve these modifications. Sophia calls this evolved mammal a lemur. Their efforts are accomplished near the middle of the Eocene Epoch, very late December 30.

Elohim and Sophia take Teams Two and Three back to the past to observe the whole Eocene Epoch. As this epoch proceeds, the climate is growing warmer. The north and south poles are temperate. Forests cover the Earth due to warm temperatures. These forests and grasslands are populated with small reptiles, early rodents, marsupials, large crocodiles, primitive hoofed animals, pythons, and turtles. The team members visit atom families that have recently evolved from large land mammals, whales and dolphins, which then migrate back to the sea. They discuss their process in evolving these creatures. Toward the end of the epoch, about 34 mya, or early December 31, they observe a sudden global warming associated with changes in oceanic and atmospheric circulation. The temperature rises six degrees centigrade for a period of twenty thousand years. This leads to a marine extinction when the deep waters became anoxic. Coral and deep-water fish are dissimated, while on land, this super-heated temperature causes mammalian life to proliferate and spread around the continents. They visit the atom communities in the magma underneath the plates of the Earth, the areas around volcanic activity, and areas where methane is being released, which is a major cause of the extinction of marine species. Team members discuss some of the causes of the extinction.

With the coming of the Oliogocene Epoch (34–23 mya) on the African continent, Team One is busy living in a lemur. They are laboring together with the other atom communities to evolve and create a new form of offspring. Cooperating in this effort and utilizing the tool-kit genes, they create a primitive mammal that Ruah names a monkey.

Ruah takes Teams One and Four on a journey to observe this epoch about 28 mya, or early morning of December 31. They discover that the continents continue to drift farther apart. The African continent moves closer to Europe, while the North American continent finally disconnects from Europe. Ice begins to cover the south pole, but not the north pole. There is a slow global cooling taking place after the Eocene extinction. Flowers and plant life continue to diversify. Woodlands are filled with new species of trees. Marine and terrestrial vertebrate fauna slowly evolve to take on a look much closer to their modern forms of life. Team members visit the atom communities involved in some of these creations and discuss methods of evolving new species.

The Miocene Epoch (23–5 mya) ushers in new creativity. In the beginning of the Miocene Epoch (23 mya, or noon on December 31), dwelling on the continent of Africa, Team One lives in a well-developed species of monkey. They are working to evolve a new species from the monkey. Their new creation would possess some of the aptitudes of the monkey, but would be larger, bigger-brained, and more land-based than the monkey. They are successful in their efforts with the help of the tool-kit genes. Ruah names this new form an ape.

About 10 mya, or early evening of December 31, Team Four is living in an ape in Africa. They decide to try to evolve a new species from their host. Amidst some personal conflict, they evolve a new mammalian species, which Elohim calls a gorilla. It is much larger than the ape and has a bigger brain, with greater hand agility.

About this same time in Africa, Team Three lives also in an ape. Along with the atom community there, Team Three organizes an effort to evolve a new species of mammal. They are finally able to create this new species, which Dabar calls a chimpanzee.

The crashing of tectonic plates against each other causes new chains of mountains to spring up. Africa and Asia grow closer together, allowing migration of various populations of animals. Receding sea levels are caused by the growing ice caps on the north and south poles. The cooling of the northern continents causes migration of some mammals to a warmer southern climate. Grasslands expand and forests recede due to a cooler, drier climate. The team members observe the results of their creative efforts to evolve and create new species of mammals, gorillas and chimpanzees. They see the results of other creative activity by atom communities in the presence of modern cats and dogs, elephants, orangutans, a surge of grazing animals, cows, sheep, pigs, wolves, horses, beavers, deer, camels, and many new species of birds. They talk to members of these communities to share stories of their creativity.

At the end of the Miocene Epoch and the beginning of the Pliocene Epoch (5–1.7 mya) in Africa, Teams Three and Four are living in a member of the chimpanzee family. A heated discussion ensues between the team members as to how the evolution will proceed. They decide that the most important change in this new species is a larger-sized brain. About 5 mya, or the middle of the evening of December 31, they evolve the first bigger-brained creature. Sophia calls this new being a hominid (its technical name is australopithecus afarensis.) This is an important event in evolutionary history

In the middle of the Pliocene Epoch (3 mya, or ten o'clock in the evening of December 31), Team One lives in a hominid on the African continent. They work to evolve and create a more developed hominid. After much effort, trial, and error, they evolve a more adaptable hominid. Elohim names this new creature homo habilis. A million year later (2 mya, or ten-thirty in the evening

336

of December 31), Team Two lives in a member of homo habilis. Again, using their creativity and tool-kit genes, they evolve a more developed hominid. Sophia calls this new creature homo erectus

The Pleistocene Epoch (1.7 mya–10,000 ya) dawns during an exciting period of evolution. About 200,000 ya, or ten minutes before midnight December 31, Teams One and Two attempt to improve the abilities of a homo erectus hominid in which they live. They evolve a more talented, larger-brained hominid, which Dabar calls archaic homo sapien.

About 50,000 ya, or two minutes before midnight December 31, Ruah takes all the team members back to the past to view this Pliocene Epoch. At the beginning of this epoch, the continental plates of the North American and South American continents finally join together, allowing the animal species to move back and forth. The whole epoch is a time of continual glacial periods, followed by glacial receding. During this epoch, there occurs the extinction of large mammals, mammoths, mastodons, saber-toothed cats, and ground sloths. In this period, flora and fauna are closely related in size and appearance to modern species.

The atom characters observe and discuss the lives and actions of their hominid creations.

They meet atom families that have been involved in various forms of creativity, and they share stories. They meet the atom families involved in the creation and evolution of homo robustus and Neanderthal homonids. They discuss the reasons for the failure of some of these creatures to survive into the present time

Teams One, Two, Three, and Four are all living in an archaic homo sapien hominid. Through observation of their hominid creations and much discussion among themselves, energized by Ruah's love energy, they set out to evolve a wiser, more intelligent hominid from archaic homo sapiens. About 50,000 ya, or two minutes before midnight December 31, divinely energized, but not directed by Ruah, they evolve homo sapiens. When they meet members of other atom families in their vicinity, they discover that the latter have also evolved homo sapiens

It is August 1, 2009, or in universal space-time, midnight December 31. Elohim, Sophia, Dabar, and Ruah gather the four teams for a journey back to the past. The Divine Beings lavishly praise and compliment the team members for their spirit of perseverance through the 15 billion years of creation history. In addition, they wish to inform their atom friends what has happened to their new species, homo sapiens, over the last fifty thousand years, or the past two minutes of universal space-time.

As they begin their journey, the Divine Beings bring them to the continent of Africa, where the atom friends have been involved in evolving hominids from the chimpanzee species over the past 5 by, and ultimately, 50,000 ya, evolving homo sapiens. Dabar begins the tour by letting his atom friends view closely the growth and propagation of homo sapiens. Dabar gives a name of homo sapiens, either *human* or *human being*. He points out that humans continue to use the primitive tools of their hominid ancestors.

Gradually, they see humans developing new tools, spears, bows, and arrows. Their ancestors had already mastered the art of fire-making. Sophia shows them how humans now live in small groups in the eastern and northeastern part of Africa. The atom friends are surprised to learn that their human creations begin to create language in the form of vocal clicks. Sophia further points out that language helps them to develop a sense of community and cooperation among themselves. She also allows them to observe how humans are evolving in consciousness and self-consciousness.

Elohim recounts how, after the development of language and then music, small bands of humans start to travel north from

Africa to the Middle East, about 30,000 ya. They observe these migratory groups establish themselves there and grow in numbers. Then Elohim traces the travels of small groups of humans who go north to Europe and south along the coast of Arabia. Over the ensuing years, the atom characters are surprised by the bravery of small groups who venture east from Arabia all the way to India and beyond, to China and Australia, and from India north to Central Asia.

Dabar informs them that, until 15,000 ya, only North and South America are devoid of human beings due to being cut off from Europe, Africa, and Asia 220 mya. In the North and South American continents, the atom friends become present to the end of the ice age. As the ice recedes, they see humans from Central Asia move east and enter the North American continent over a land bridge at the Bering Strait. They advance from north to south into the South American continent. By 12,000 ya, they are shocked to see the human race reside on all of the continents of planet Earth.

About 10,000 ya, the atom friends are astonished by the ability of humans to begin to domesticate animals and grain crops. Their advances in language skills allow them to begin to organize themselves into small towns, they discover. Commerce between human communities continues to expand. Dabar observes that written language is developed with the invention of an alphabet in the Middle East and Sanskrit in India. By 4,000 ya, the world population is 5 to 10 million.

As the human population grows, Ruah shows her atom friends how humans are organizing cities. These communities become places of growing civilization, she tells them. Further, many of these cities and surrounding territories become nations ruled by kings. Cities become places of power, spheres of influence, commerce, financially supporting themselves through a system of taxation, and building armies to protect their people and way of life. This leads to continual wars, with one nation trying to conquer another to expand its power, wealth, and influence, Ruah concludes.

Ruah highlights the development of human civilizations. She shows them the ancient civilizations of Sumeria, Egypt, Israel, and later, Greece. Starting about 2,600 ya, Ruah takes them around the Earth to show where humans are developing religions in a search for gods or a God who exists invisibly beyond the Earth, but relates intimately with Earth-bound humans. Some religions search for a life after death in the presence of their God. The atom friends are shown the areas where humans have begun Confucianism, Buddhism, Hinduism, Zoroastrianism, and other lesser beliefs. About 2,000 ya, Christianity develops in the Middle East and spreads around the known world, while 1,400 ya, the Muslim religion comes into existence in the Middle East and spreads out from there, Ruah tells them. For about a thousand years, there are wars between Christians and Muslims, especially for control of the city of Jerusalem in the Middle East.

Human knowledge grows through the centuries in the areas of philosophy, theology, cosmology, physics, biology, chemistry, mathematics, and astronomy. As they visit the nations of Europe from 1,400 ya up to the present, they are mystified that their human creations seem to be involved in endless wars while utilizing more and more developed and deadly war technologies. Why are their self-conscious, intelligent, and sensitive creations killing each other? they inquire of Elohim. They realize that the purpose of much of their warfare has to do with the acquisition of power, wealth, commerce, and influence. Elohim takes them to Europe about 300 ya, where they observe the beginnings of the Industrial Revolution. He shows them the industrial progress during the ensuing three hundred years.

Finally, Elohim takes them to view some of the astounding human developments of the last hundred years: the invention of the car and airplane; the harnessing of the energy in the atom and development of atomic and nuclear bombs; the computer; the copier; medical technologies; space travel; globalization; the tremendous growth of the human population; the ecological devastation and pollution of the oceans, rivers, land, and atmosphere; the destruction of many species of flora and fauna by unwise human habits; growth of the world's education systems; and the persistence of religions and religious belief in gods or God.

Sophia concludes their journey with this question: "My atom friends, is your creative mission at an end, or is their more evolution and creativity to come as you continue to live in space-time on planet Earth in this solar system?

Book II contains a surprise ending. Unbeknown to the twelve atom characters, they all end up in me, the visionary, and in my body-person. I hold a conversation with them in which I inform them that I had been present and followed them closely during all the 15 billion years of their creative and evolutionary journey. We address Sophia's concluding question, discuss their ongoing relationship with Elohim, Sophia, Dabar, and Ruah, the possible future of the human race, and the ultimate destiny of planet Earth and the universe!

Sources Consulted

Apfel, Mecia H. *Nebulae*. New York: Lothrop, Lee & Shepard Books, 1988.

Arthur, Wallace. *The Origin of Animal Body Plans*. New York: Cambridge University Press, 1997.

Asimov, Isaac. *The Exploding Suns*. New York: Dutton, 1985.

Asimov, Isaac. *Beginnings: The Story of Origins*. New York: Walker, 1987.

Asimov, Isaac. *Our Solar System*. Milwaukee, WI: Gareth Stevens Pub., 1988.

Asimov, Isaac. *Atom*. New York: Dutton, 1991.

Bailey, Jill. *Evolution and Genetics*. New York: Oxford University Press, 1995.

Bailey, Lloyd R. *Genesis, Creation and Creationism*. New York: Paulist Press, 1993.

Ball, Philip. *Life's Matrix*. New York: Farrar, Straus and Giroux, 2000.

Barbree, Jay. *A Journey Through Time*. New York: Penguin Studio, 1995.

Barrow, John D. *The Origin of the Universe*. New York: Basic Books, 1994.

Bartusiak, Marcia. *Through a Universe Darkly*. New York: HarperCollins Publishers, 1993.

Bates, Marston. *The Forest and the Sea*. Chicago: Time-Life Books Inc., 1960.

Begelman, Mitchell. *Turn Right at Orion*. Cambridge, MA: Helix Books-Perseus Publishing, 2000.

Behe, Michael J. *Darwin's Black Box*. New York: Touchstone Books—Simon & Schuster, 1996.

Boslough, John. *Stephen Hawking's Universe*. New York: Avon Books, Division of Hearst Corp., 1985.

Branley, Franklin M. *The Sun and the Solar System*. New York: Twenty-First Century Books, 1996.

Branley, Franklin M. *The Nine Planets*. New York: Crowell, 1978.

Branley, Franklin M. *Journey into a Black Hole*. New York: T. W. Crowell, 1986.

Brown, Lester R., Flavin, Christopher, and Postel, Sandra. *Saving the Planet*. New York: W. W. Norton & Co., 1991.

Bruteau, Beatrice. *God's Ecstasy*. New York: The Crossroad Publishing Co., 1997.

Calder, Nigel. *Einstein's Universe*. New York: Penguin Books, 1979.

Carson, Rachel. *The Sea Around Us*. New York: Signet Science Library Book published by The New American Library, 1961.

Carson, Rachel. *Silent Spring*. Boston, MA: Houghton Mifflin Co., 1962.

Chaisson, Eric J. *Cosmic Evolution*. Cambridge, MA: Harvard University Press, 2001.

Chown, Marcus. *Afterglow of Creation*. Sausalito, CA: University Science Books, 1996.

Clark, Stuart. *Stars and Atoms*. New York: Oxford University Press, 1995.

Consolmagno, Br. Guy SJ. *The Way to the Dwelling of Light*. Rome, Italy: Vatican Observatory, 1998.

Conway, Morris S. *The Crucible of Creation*. New York: Oxford University Press, 1998.

Croswell, Ken. *The Alchemy of the Heavens*. New York: Anchor Books, 1995.

Darling, David J. *The Galaxies: Cities of Stars*. Minneapolis, MN: Dillon Press, 1985.

Darling, David J. *Deep Time.*, New York: Delacorte Press, 1989.

deChardin, Teilhard. *The Future of Man*. New York: Harper and Row, Publishers, 1939.

deChardin, Teilhard. *The Phenomenon of Man*. New York: Harper and Row, Publishers, 1955.

deChardin, Teilhard. *The Appearance of Man*. New York: Harper and Row, Publishers, 1955.

deChardin, Teilhard. *The Divine Milieu*. New York: Harper and Row, Publishers, 1957.

deChardin, Teilhard. *Building The Earth*. Wilkes-Barre, PA: Dimension Books, 1965.

deDuve, Christian. *Vital Dust,* New York: Basic Books, 1995.

deDuve, Christian. *Life Evolving.* New York: Oxford University Press, 2002.

Dickinson, Terrence. *The Universe and Beyond,* 3rd Edition. Willowdale, Ontario: Firefly Books Ltd., 1999.

Dyson, Freeman. *Origins of Life,* 2nd Edition. New York: Cambridge University Press, 1999.

Eicher, David J. *Beyond the Solar System.* Waukesha, WI: Kalmbach Books, 1992.

Eiseley, Loren. *The Night Company.* New York: Charles Scribner's Sons, 1947.

Eiseley, Loren. *The Firmament of Time.* New York: Atheneum, MacMillan Publishing Co., 1960.

Ellis, George F. R. *Before the Beginning.* New York: Bjoyars/Bowerdean, 1993.

Engelbert, Phyllis, and Dupuis, Diane L. *The Handy Space Answer Book.* Canton, MI: Visible Ink Press, 1998.

Erickson, Jon. *Plate Tectonics.* New York: Facts on File, 1992.

Faricy, Robert S. J. *The Spirituality of Teilhard de Chardin.* Minneapolis, MN: Winston Press, 1981.

Feinberg, Gerald. *What Is the World Made of?* 1st Edition. Garden City, NJ: Anchor Press/Doubleday, 1977.

Ferris, Timothy. *The Whole Shebang.* New York: Simon & Schuster, 1997.

Ferris, Timothy. *Coming of Age in the Milky Way.* New York: Morrow, 1998.

Fodor, R. W. *The Strange World of Deep Sea Vents*. Hillside, NJ: Enslow, 1991.

Ford, Adam. *Universe*. Mystic, CT: Twenty-Third Publications, 1987.

Fortney, Richard. *Life*. New York: Alfred A. Knopf, 1998.

Fox, Matthew. *Original Blessing*. Santa Fe, NM: Bear & Company, 1983.

Fox, Matthew. *The Coming of the Cosmic Christ*. San Francisco, CA: Harper and Row Publishers, 1988.

Gallant, Roy A. *Comets, Asteroids and Meteorites*. Tarrytown, NY: Benchmark Books, 2001.

Gleiser, Marcelo. *The Dancing Universe*. New York: Dutton, 1997.

Goldsmith, Donald. *Mysteries of the Milky Way*. Chicago: Contemporary Book, 1991.

Goodwin, Simon. *Hubble's Universe*. New York: Penguin Studio–Viking Penguin, 1996.

Gould, Stephen J. *Full House*. New York: Three Rivers Press, 1996.

Greene, Brian. *The Elegant Universe*. New York: Vintage Books–Random House Inc., 1999.

Gribbin, John R. *Fire on Earth*. New York: St. Martin's Press, 1996.

Guth, Alan H. *The Inflationary Universe*. Reading, MA: Addison-Wesley Pub., 1997.

Haught, John F. *Deeper Than Darwin*. Cambridge, MA: Westview, A Member of Perseus Books Group, 2003.

Haught, John F. *Christianity and Science*. Maryknoll, NY: Orbis Books, 2007.

Hays, Edward. *Prayers for a Planetary Pilgrim*. Leavenworth, KS: Forest of Peace Books, 1989.

Hitchcock, John. *The Web of the Universe*. New York: Paulist Press, 1991.

Holland, Heinrich D. *The Chemical Evolution of the Atmosphere and Oceans*. Princeton, NJ: University Press, 1984.

Hughes, David. *Story of the Universe*. Mahwah, New Jersey: Troll Associates, Eagle Book Limited, 1991.

Jesperson, James, and Fitz-Randolph, Jane. *Looking at the Invisible Universe*. New York: Atheneum, 1990.

Johnson, Elizabeth A. *Quest for the Living God*. New York: The Continuum International Publishing Group Inc., 2007.

Joseph, Lawrence E. *Gaia: The Growth of an Idea*. New York: St. Martin's Press, 1990.

Kerrod, Robin. *The Solar System*. Minneapolis, MN: Lerner Publications Co., 2000.

Krauss, Lawrence M. *Atom*. Boston, MA: First Back Bay–Little, Brown and Company, 2002.

Kunzig, Robert. *The Restless Sea*. New York: Norton, 1999.

Lederman, Leon M. *From Quarks to the Cosmos*. New York: Scientific American Library, Distributed by W. H. Freeman, 1989.

Levi-Setti, Riccardo. *Trilobites.* Chicago: University of Chicago Press, 1993.

Levy, Charles K. *Evolutionary Wars.* New York: W. H. Freeman and Co., 1999.

Lewin, Roger. *Thread of Life.* New York: Smithsonian Books, Distributed by Norton, 1982.

Liderbach, Daniel. *The Numinous Universe.* New York: Paulist Press, 1989.

Lonergan, Anne, Edited by Richards, Caroline. *Thomas Berry and the New Cosmology.* Mystic, CT: Twenty-Third Publications, 1987.

Lovelock, James. *The Age of Gaia.* New York: Norton, 1988.

Margulis, Lynn. *Symbiosis in Cell Evolution,* 2nd Edition. New York: W. H. Freeman and Company, 1981, 1993.

Margulis, Lynn, and Sagan, Dorion. *Microcosmos.* Berkley, CA: University of California Press, 1986.

Margulis, Lynn, and Sagan, Dorion. *What is Life?* New York: Simon & Schuster, 1995.

Margulis, Lynn. *Symbiotic Planet.* New York: Basic Books–The Perseus Books Group, 1998.

Margulis, Lynn, and Sagan, Dorion. *Acquiring Genomes.* New York: Basic Books–The Perseus Books Group, 2002.

Marshall, Elizabeth L. *The Human Genome Project.* New York: F. Watts, 1996.

Mather, John C., and Boslough, John. *The Very First Light.* New York: Basic Books, 1996.

McGaa, Ed (Eagle Man). *Mother Earth Spirituality*. San Francisco, CA: Harper San Francisco–A Division of HarperCollins Publishers, 1990.

McSween, Harry Y. Jr. *Meteorites and Their Parent Planets*. New York: Cambridge University Press, 1987.

McSween, Harry Y. Jr. *Fanfare for Earth*. New York: St. Martin's Press, 1997.

Michod, Richard E. *Eros and Evolution*. Reading, MA: Addison-Wesley Pub. Co., 1995.

Miller, Russell. *Continents in Collision*. Alexandria, VA: Time-Life Books, 1983.

Moore, Patrick. *The Universe for the Beginner*. New York: Press Syndicate of University of Cambridge, 1990.

Morris, Richard. *Cosmic Questions*. New York: Wiley, 1993.

Newberg, Andrew, D'Aquili, Eugene, and Rause, Vince. *Why God Won't Go Away*. New York: Ballantine Books, 2001.

Nicholson, Cynthia P. and Slavin, Bill. *Discover the Planets*. Toronto, Canada: Kids Can Press Ltd., 1998.

Norman, David. *Prehistoric Life*. New York: MacMillan, 1994.

Olomucki, Martin. *The Chemistry of Life*. New York: McGraw-Hill, 1993.

Olson, Steve. *Mapping Human History*. New York: Houghton Mifflin Company, 2002.

O'Murchu, Diarmuid. *Quantum Theology*. New York: Crossroad Publishing Company, 1997.

O'Murchu, Diarmuid. *Evolutionary Faith*. Maryknoll, NY: Orbis Books, 2002.

Parker, Barry, R. *Colliding Galaxies*. New York: Plenum Press, 1990.

Peters, Ted, Editor. *Cosmos as Creation*. Nashville, TN: Abingdon Press, 1989.

Polkinghorne, John. *The Faith of the Physicist*. Princeton, NJ: Princeton University Press, 1994.

Prager, Ellen J., with Earle, Sylvia A. *The Oceans*. New York: McGraw-Hill, 2000.

Raven, Peter H., and Johnson, George B. *Biology*, 4th Edition. New York: WCB McGraw-Hill, 1996.

Roberts, Elizabeth, Editor, and Amidon, Elias. *Earth Prayers from around the World*. San Francisco, CA: Harper San Francisco, A Division of HarperCollins Publishers, 1991.

Ronan, Colin A. *The Natural History of Universe*. New York: MacMillan, 1991.

Ronan, Colin A., Editor. *The Universe Explained*. New York: H. Holt, 1994.

Rosen, Sidney. *Which Way to the Milky Way?* Minneapolis, MN: Carolhoda Books, 1992.

Roszak, Theodore. *The Voice of the Earth*. New York: Simon & Schuster, 1992.

Rowan-Robinson, Michael. *Ripples in the Cosmos*. New York: W. H. Freeman Spektrum, 1993.

Ruether, Rosemary R. *Gaia and God*. San Francisco, CA: Harper San Francisco–A Division of HarperCollins Publishers, 1992.

Sagan, Carl. *Cosmos*. New York: Random House, 1988.

Savage, Robert J. G. *Mammal Evolution*. New York: Facts on File, 1986.

Schroeder, Gerald L. *The Hidden Face of God*. New York: Touchstone Book, Simon & Schuster, 2001.

Simon, Seymour. *Galaxies*. New York: Morrow Junior Books, 1988.

Simon, Seymour. *Our Solar System*. New York: Morrow Junior Books, 1992.

Sipiera, Paul P. *The Solar System*. New York: Children's Press, 1997.

Smoot, George, and Levin, Roger. *Wrinkles in Time*. New York: W. Morrow, 1993.

Southwood, Richard. *The Story of Life*. New York: Oxford University Press, 2003.

Stanley, Steven M. *Extinction*. New York: Scientific American Library, Distributed by W. H. Freeman, 1987.

Starhawk, *The Spiral Dance*. San Francisco, CA: Harper San Francisco–A Division of Harper Collins Publishers, 1979.

Svarney, Thomas E., and Barnes-Svarney, Patricia. *The Handy Ocean Answer Book*. Farmington Hills, MI: Visible Ink Press: A Division of Gale Group, Inc., 2000.

Swimme, Brian, and Berry, Thomas. *The Universe Story*, San Francisco, CA: Harper, 1992.

Swimme, Brian. *The Hidden Heart of the Cosmos*. Maryknoll, NY: Orbis Books, 1996.

Thomas, Lewis. *The Lives of a Cell*. New York: Bantom Books–Viking Penguin, 1974.

Tilby, Angela. *Soul—God, Self and the New Cosmology*. New York: Doubleday, 1992.

Tilling, Robert I. *Born of Fire*. Hillside, NJ: Enslow Publishers, 1991.

Trefil, James. *The Moment of Creation*. New York: Scribner, 1983.

Trefil, James, and Hazen, Robert M. *The Sciences*, 2nd Edition. New York: John Wiley & Sons, Inc., 1998.

Tucker, Wallace H. *The Dark Matter*. New York: Morrow, 1988.

Wald, Robert M. *Space, Time, and Gravity*. Chicago: University of Chicago Press, 1992.

Walker, Gabrielle. *Snowball Earth*. New York: Crown Publishers, 2003.

Ward, Peter D., and Brownlee, Donald. *Rare Earth*. New York: Copericus Springer-Verlag New York Inc., 2000.

Werblowsky, R. J. Zwi, and Wigoder, Geoffrey, Editors. *The Oxford Dictionary of the Jewish Religion*. New York/Oxford: Oxford University Press, 1997.

Wessels, Cletus. *The Holy Web*. Maryknoll, NY: Orbis Books, 2000.

Wolpert, Lewis. *Principles of Development*, 2nd Edition. New York: Oxford University Press, 2002.

Young, Louise B. *The Unfinished Universe*. New York: Oxford University Press, 1986.

Zimmer, Carl. *Evolution*. New York: HarperCollins Publishers, 2001.

Glossary

ADP (adenosine diphosphate). A nucleotide made up of adenine, sugar, and two phosphate groups that is formed from ATP by the removal of one phosphate group, which then releases the usable energy to a cell.

ATP (adenosine triphosphate). A nucleotide consisting of adenine, sugar, and three phosphate groups that captures light energy in the process of photosynthesis. It later loses one phosphate group to become ADP.

amino acid. A building block of proteins. There are twenty different amino acids, which allows for the building of hundreds of combinations of proteins.

atoms. The fundamental building blocks for all matter. Each atom consists of a positively charged nucleus and negatively charged electrons, which orbit the nucleus. There are ninety-two different atoms.

life atoms. Five atoms: carbon, oxygen, hydrogen, nitrogen, and phosphorus were involved in the beginning of biological life on planet Earth about 3.6 billion years ago. They are still intimately involved in the continuation of biological life on Earth. A sixth atom, sulfur, was involved with the other five atoms in the beginning and continuation of biological life near the volcanic vents on the floor of the oceans of planet Earth.

autopoesis. The ability of atoms, photons, molecules, and other substances to act creatively.

biological life. The ability of some individual beings to be organized in such a manner that they are able to sustain themselves through nourishment, to grow and live, and then to reproduce copies of themselves that are able to live, sustain, grow, and reproduce offspring.

black hole. An object formed at the death of some very, very large suns. Its mass is so dense and concentrated that little or nothing—not even light—can escape its surface.

body plan. The entire design plan that DNA in bacteria and cells have encoded on themselves, used in the building and reproducing of an offspring.

centriole. An organelle that organizes and divides the spindle fibers during the reproduction process.

DNA (deoxyribonucleic acid). A strand of nucleotide made up of a phosphate, sugar, and base groups. There are five base groups, which allows for many different DNA combinations. DNA uses four of the base groups: T-base (thiamine), A-base (adenine), G-base (guanine), and C-base (cytosine) in building its genetic code.

electromagnetic force. One of the four forces of nature in our universe. It refers to the unified nature of electricity and magnetism, which gives rise to different forms of visible and invisible radiation of various frequencies.

eukaryote cell. A small organism that contains its double-helix DNA within a membrane-bound nucleus. All plant, animal, human, and fungal life is made up of multicellular eukaryote cells. This cell evolved by the symbiosis of several species of prokaryote bacteria.

flagellum. A lashlike appendage that helps certain bacterial species to move from place to place.

four fundamental forces. There are four fundamental forces in the universe—gravity, strong force, electromagnetism, and weak force. It is theorized that these four forces were unified at the first moment of the creation of the universe and split off as the latter expanded.

galaxy. A large system of hundreds of billions of suns (stars), nebulae, and other celestial bodies. They appear like large cities of hundreds of billions of lights in the darkness of space-time. There are different types of galaxies: spiral, barred spiral, elliptical, ring, and irregular.

gravity. One of the four forces of nature in our universe. It is an attractive force that acts on every object, large and small, in our universe. It creates a web that draws and holds all objects in relation to each other.

helium nuclei. An atom of helium that has lost its orbiting electrons.

hydrogen atom. The first atom to come into existence at the beginning of the universe.

kinetochore. Disk-shaped structures to which the spindle fibers are attached at both ends of the spindle apparatus.

macrocosm. Denotes the whole of our universe—all of its billions of galaxies, trillions of suns (stars), and black empty space in between the galaxies.

microcosm. The miniature, microscopic world, as contrasted to the macrocosm, where tiny atoms, molecules, bacteria, and other small beings exist, live, and operate.

microtubule. A tiny, long, and hollow protein cylinder that provides the functional internal structure of the flagella in bacteria.

mitochrondia. The powerhouse for the bacterium or cell. It is a sausage-shaped organelle within the bacterium or cell where molecules derived from sugars react with oxygen to produce energy for the bacterium or cell to live, grow, and reproduce.

molecule. A cluster of atoms bound together to form different kinds of material.

nebulae. Gigantic clouds of mainly hydrogen and helium gases, located in all of the galaxies in the universe. They are the nurseries where suns (stars) are birthed.

neutron. One of the two building blocks of the nucleus of an atom. It carries no electrical charge. It is made up of three quarks.

photon. This is a particlelike unit of light that lacks any mass. They were birthed in the first moment of the creation of the universe. It is a single packet of electromagnetic energy or radiation.

photosynthesis. The mechanism within bacteria or cells by which the photonic light of the sun is converted into energy to help them live, grow, and reproduce on planet Earth.

prokaryote bacterium. A primitive bacterium that lacks a membrane-bound nucleus within which the DNA helix lives. Biological life began with the creation of prokaryote bacteria about 3.6 billion years ago.

protein. This is an extremely complex molecule, typically consisting of more than one hundred amino acids and millions of atoms. Proteins direct the chemical activity within bacteria and eukaryote cells.

proton. One of the two building blocks of the nucleus of an atom. Protons are positively charged and consist of three quarks.

quark. It is the fundamental building block of protons and neutrons, which are made up of three quarks.

RNA (ribonucleic acid). A strand of nucleotide made up of a phosphate, sugar, and base group. Instead of the T-base (thiamine), it uses the U-base (uracil) along with the A-base (adenine), G-base (guanine), and C-base (cytosine) in building its genetic code. RNA is the copier of the DNA body plan, the transfer messenger of the plan, and the builder of the new offspring. In the beginning of bacterial life 3.6 billion years ago, RNA carried out two functions: first, it was encoded with the body plan of the bacterium, and second, it built the new offspring using the plan. Later, DNA took over the first function from RNA, which retained the second function.

solar system. This is a system located on the Cyrus-Orion arm of the Milky Way galaxy, which is made up of our Sun, which is orbited by nine planets, including our planet Earth.

spindle apparatus. This is an assembly within a bacterium or eukaryote cell made up of the spindle fibers and held in place by the disklike kinetochore. It is the location where the division of cells took place when multicellular life began.

strong force. One of the four fundamental forces in nature. It is the force that holds the nucleus of an atom (the quarks within the protons and neutrons) together. It operates over very small distances.

sun (star). A body that contains hydrogen and helium nuclei gases. At its core is the nuclear furnace, which is fueled by the helium nuclei. It radiates various forms of photons that energize the solar system. In looking up into the sky and universe, we call suns, stars.

stromatolite. A mat of ancient bacteria formed 2 to 3 billion years ago. The different species of bacteria lived a community life above the level of the ancient sea. The fossilized mats can still be seen at various parts of the seashores on planet Earth today.

supernova. This is the stupendous explosion of an extremely large sun, which increases its brightness hundreds of millions of times in a few days. The implosion and explosion of a supernova cause the star to die but give birth to all of the ninety-two chemical elements that seed the nearby space-time environment.

symbiosis. The process whereby two or more dissimilar bacteria live together in a close association. It is believed that the eukaryote cell came into existence by the process of symbiosis, bringing improvements to its larger host by sharing its DNA with the latter.

weak force. One of the fundamental forces in nature. It is the force involved in reactions within atomic structures, such as beta decay.

Made in the USA
Lexington, KY
11 October 2016